MATERIALS SCIENCE AND TECHNOLOGIES

DISPERSION DYNAMICS IN THE HALL EFFECT AND PAIR BONDS IN HITC

MATERIALS SCIENCE AND TECHNOLOGIES

Additional books in this series can be found on Nova's website under the Series tab.

Additional e-books in this series can be found on Nova's website under the eBooks tab.

MATERIALS SCIENCE AND TECHNOLOGIES

DISPERSION DYNAMICS IN THE HALL EFFECT AND PAIR BONDS IN HITC

ANTONY J. BOURDILLON

Library of Congress Cataloging-in-Publication Data

ISBN: 978-1-53612-568-9

Published by Nova Science Publishers, Inc. † New York

For Alice

CONTENTS

PREFACE

The conundrum is this:

What differentiates high temperature superconductivity from low temperature superconductivity?

Hi*Tc* has the positive Hall coefficient, $R_H > 0$.
How come?
Its charge carriers are holes.
What is a hole?
An electron vacancy in an otherwise filled valence band.
Can the Lorentz force act on a void?
No!
How can R_H be positive?
Consider first the free particle…

This book is a sequel to our earlier description of empirical processing of cuprate, high temperature superconductors (HiT_c)[1,2]. Chemical 'holes'

[1] This T_c represents critical superconducting temperature and need not be confused with the Curie temperature in ferromagnetic materials.

[2] Bourdillon A and Tan-Bourdillon N.X., *High Temperature Superconductors: Processing and Science*, Academic Press, N.Y., 1994, ISBN 0-12-117680-0.

are critical, and the fact is widely acknowledged. Measurement of associated charge carriers in electrical conductivity is therefore indispensable. There are anomalies in the traditional understanding of the operation of the Lorentz force on these carriers: the force cannot influence voids. However, by basing an argument on dispersion dynamics of free particles, we find a consistent solution for the Hall effect measurements that are made. The developments follow from applications of the stable wave packet to the elements of quantum mechanics. The dynamics are adapted from free particles, to particles in crystal fields. The method is so transparent that any experimental physicist can grasp its significance for the work that has to guide further research in the field. In particular, the dynamics are demarcated not by positive or negative charge; but by effective mass. This is negative when the curvature of the electronic energy bands are negative, or positive when positive. Moreover, all of the positive nuclear charges are immobile so that the Lorentz force influencess only negatively charged electrons as charge carriers.

The cuprate superconductors were discovered thirty years ago. After describing the processing characteristics, we left the full explanation to a later time. Meantime, multiple partial theories have been proposed. There remained a need for a unified theory that nevertheless differentiated conventional low temperature alloy superconductors (LoT_c), from the newer ceramic HiT_c compounds.

The model of the wave group is sufficiently fundamental and definite that it has enabled new understanding, not only in the electrodynamics of the Hall effect; but also in the combined mechanisms for electrical conductivity at normal temperature, and for superconductivity at lower temperatures. In particular, a formula is developed that describes acceleration of charge carriers in a crystalline lattice, whether it is that of a metal; or of a doped semiconductor; or of a ceramic superconductor. The chemically derived 'hole' states and Fermionic charge carriers, that are measured and characterized in normal conductivity at raised temperatures $T > T_c$, attract at lower temperatures Boson electron pairs that travel through the extensive crystal lattice without resistance. We understand that, in HiT_c , the chemical 'holes' induce excitons that replace the isotope effect in the BCS theory of LoT_c [3].

"In questions of science, the authority of a thousand is not worth the humble reasoning of a single individual" [4]. It is therefore not so significant that the topics discussed in this book have, for the most part, previously passed muster in disparate refereed journals; more valuable are the consistent, collated reasons now presented in a unified whole. In particular, research papers suffer space restrictions that are relieved in this expansion of a recent solution [5] for high temperature superconductivity. Some readers, depending on particular interest, may prefer to approach the chapters in reverse order. Their individual contents are sufficiently diverse to support this rearrangement; though the presentation in this book requires the logical order adopted. Appendices are designed for added clarity, with tutorial purpose.

The American Physical Society March Meeting recently bade the author, "Be aggressive; shake off the kid gloves; add insult." I do humble reasoning.

AJB.
San Jose, California.

Cover page: Functions of free relativistic particles.

---

[3] Bardeen J., Cooper L.N. and Schrieffer J.R., 1957, *Phys. Rev.* 108 1175
[4] Galilei G. 1632,
[5] Bourdillon A.J.,2017, *Journal of Modern Physics* **8** 483-499,doi: https:// doi. org/ 10. 4236/ jmp.2017.84031

ACKNOWLEDGMENTS

Author has the benefit, spread over several years, of helpful criticism from the editor and referees for the *Journal of Modern Physics*.

Chapter 1

DISPERSION DYNAMICS IN FREE PARTICLES

1.1. WHAT IS QUANTUM MECHANICS?

This book is written for consistency in a semi-unified explanation for high critical temperature superconductivity (HiT_c) - semi-unified because it applies only partly to low temperature superconductivity (LoT_c), previously described in the classic work of Bardeen, Cooper and Schrieffer [1]. Their solution depends on the isotope effect (experimental T_c/M_i^2 = constant for isotopes with mass M_i, where lattice distortions enable the formation of Bosonic Cooper pairs of otherwise independent Fermionic electrons). The isotope effect is only weakly apparent in HiT_c compounds [2, 3]; the role of isotopes in LoT_c metals and alloys is replaced in HiT_c compounds, by responses to chemical 'holes.' In order to clarify the groundwork, it will help to define quantum mechanics as a practical tool and to develop the explanation for Hall transport in all of semiconductors, metals, and HiT_c compounds. Our method is empirical and locates a defined path between mathematical complexity [4, 5, 6] and contrasting simplicity in the motion of free particles.

For perspective, trace the origin of the quantum mechanics back to corpuscular theory of light and subsequent replacement by the wave mechanics of Huygens, Fraunhofer, Fresnel and others. Subsequently,

Maxwell's electromagnetism provided a deep understanding of light waves including their energy and wave properties prior to discoveries of quantization. Electromagnetism can be applied to waves of infinite extension or to the wave packets used here. Meanwhile, atomic spectra in either emission or absorption contain sharp lines that required further understanding. Examples are the series of lines in the spectra of atomic hydrogen, including those named after Lyman, Balmer, Paschen, Brackett and Pfund. Bohr ascribed the discrete lines to standing waves in his atomic model. The transition energies depended on differences between principal quantum numbers in initial n_i and final n_f states, $\sim(1/n_f^2-1/n_i^2)$. Schrödinger clinched the model by calculating the eigenvalues of his non-relativistic wave equation for the hydrogen atom. The discovery founded modern chemistry and stimulated new breakthroughs in physics: in symmetric and antisymmetric wave functions for Bosons and Fermions, obeying Bose-Einstein statistics or Fermi-Dirac statistics respectively for distinguishable and indistinguishable particles; in corresponding commuting or anti-commuting operators; in a description for the intrinsic spin of particles; in new ways for calculating creation and annihilation by these specific operators; *etc.*, in endless progression.

The Schrödinger equation, that allows solutions for standing waves in atomic structure, likewise applies to bound states in the harmonic oscillator and to wave functions in a bounded and periodic solid which can then be treated, by approximation, as infinite in extent. We understand that the wave function will be measured in an *event* by standing-wave, quantized excitations, and that it therefore appears to be quantized, but that the allowed distribution of energies in perfectly free electron motion or photon emission are continuous, as expressed through Maxwell's equations for electromagnetism.

Our purpose is not, as in elementary particle physics, a simulation of hadron-meson interaction by creation and annihilation operators, in fields subject to Hamiltonian and Lagrangian equations. Our aim is an understanding of electron transport in the crystal fields of metals or doped semiconductors, and of ceramic superconductors. We bypass Dirac's pronouncements [7] (the unstable wave packet; the electron velocity c [6], jitterbugging more and less; the positive mass of the positron; with positive kinetic energy)[7] but share his relativistic solution that gives the antiparticle a negative eigenvalue. It is very well to create a mathematical theory of quantization so long as that is not confused with the corresponding, falsifiable, physical hypothesis. Mathematicians need a beautiful theory; physicists have to feel it working. By example, our free particle model runs parallel to sophisticated treatments of the solid state that employ the packet with tensor properties [8], though the concentration here on the propagation direction leads us directly to its focused conclusion.

There are other anomalies. We all need numbers for measurement, but scientists have often crossed the line from empirical physics to creative mathematics. An example, illustrated by consequences of the stable wave packet, is Heisenberg's Uncertainty since it is often called an axiom. However, an axiomatic treatment in mathematics is typically a creative short cut for an alternative physical hypothesis that is falsifiable [9, 10]. As history shows, sometimes the hypotheses themselves are mutually contradictory, but have equal scientific standing until one or other is falsified. The following treatment shows that prior wave mechanics provides the empirical explanation for the Hall coefficient and HiT_c.

1.2. The Stable Wave Packet

The reader can easily verify that the following wave function for a free particle is stable:

$$\phi(\bar{k}, x, \bar{\omega}, t) = A.\exp\left(\frac{X^2}{2\sigma^2} + X\right) \quad \text{with} \quad X = i(\bar{k}x - \bar{\omega}t) \qquad (1.1)$$

---

[6] He recognized a difficulty. It conflicts with relativity.

[7] Wigner quipped at his brother-in-law, "There is no god and Dirac is his prophet," Times Literary Supplement April 7, 2017.

where X is an imaginary space-time variable that causes ϕ to oscillate inside a Guassian envelope (Figure 1.1). In particular, the *mean* wave vector $\overline{k}$, and *mean* angular frequency $\overline{\omega}$, are themselves stable not only because they are mean values in a symmetric wave function; but their stability is also guaranteed by respective conservation of momentum and energy; and triple guaranteed by symmetry in space-time. The wave is one-dimensional in the direction of propagation; the two transverse components are planar. Their dynamics, derived from relativity, will be described below. The coherence σ is particular because it depends on initial conditions, but it is stable during propagation in free space as a consequence of Newton's first law of motion. The variable determines the width of the wave packet, in space or time. The normalizing amplitude A depends on σ and has corresponding stability. The envelope, $\exp(X^2/2\sigma^2)$, depends on the square of imaginary X, itself a function of four variables. Two are already described, so we are left with the variables x and t that describe the profile. Since the other variables are all stable, this profile is also stable. With a stable wave group, we can progress to new physics. First consider particles; we will apply the formulae to antiparticles in the next section and, obliquely, in following chapters.

The special theory of relativity defines the relationships between angular frequency, wave vector and velocity. From Einstein's formula for the energy E of a particle of momentum p

$$E^2 = p^2c^2+m_o^2c^4 \tag{1.2}$$

where c is the speed of light. Substitute for E from Planck's law and for p from the de Broglie hypothesis, to find that:

$$\omega^2 = k^2c^2 + \left(\frac{m_0c^2}{\hbar}\right)^2 \tag{1.3}$$

and by differentiating after simplifying units, $\hbar = c = 1$:

$$\frac{\omega}{k}\frac{d\omega}{dk} = c^2 = v_p v_g \tag{1.4}$$

where ω/k is the phase velocity v_p and $d\omega/dk$ the group velocity v_g [section 1.5 below]. These are illustrated in Figures 1.2 and 1.3. For a particle, $m_o > 0$ while $v_p > c$, and $v_g < c$ as is required by relativity. The same result is obtained after operating on equation 1.1 with the relativistic version of the Klein-Gordon equation. There are many consequences, including Newtonian time for the carrier wave within the coherence of the rest frame [11]. The latter has explanatory value in the reduction of the wave packet during quantum measurement, occurring typically faster than the group velocity allows. The objectives of this book direct discussion to mass, charge, motion, force, energy and acceleration.

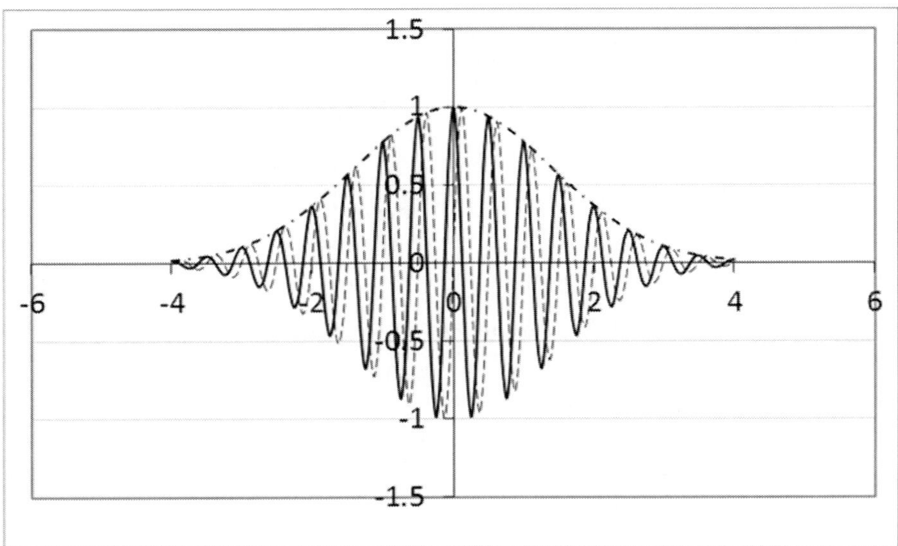

Figure 1.1. In particles with rest mass greater than zero ($m_o > 0$), the phase velocity v_p of the carrier wave containing real part (full line) and imaginary part (dashed) is greater than the group velocity v_g of the envelope (dash dot), $v_p . v_g = c^2$. In photons, $m_0 = 0$ and in free space $v_p = v_g = c$ The ordinate scale measures amplitude; the abscissa scale measures the space-time variable, x or t.

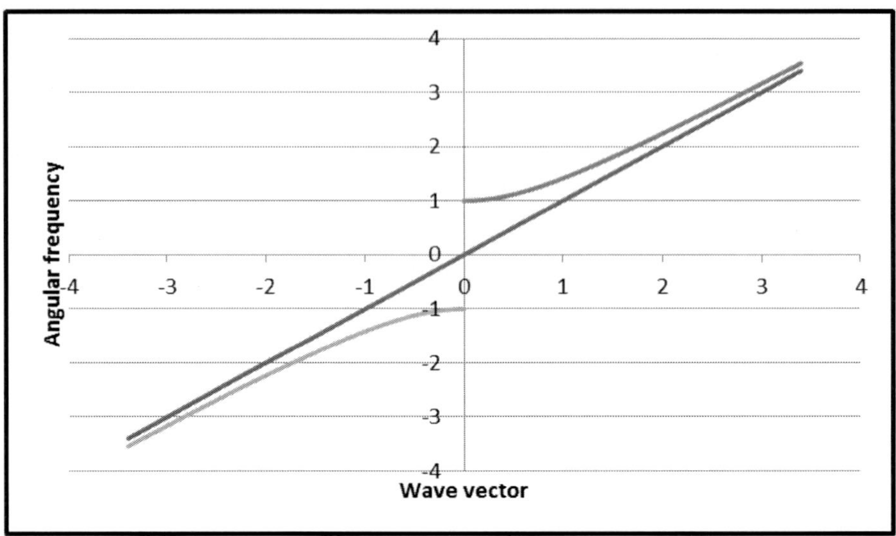

Figure 1.2. Angular frequency plotted against wave vector following equation 1.3, $\omega = \pm (k^2+m_0^2)^{1/2}$, using simplified units and a unit mass $m_0 = 1$ for the particle and $m_0 = -1$ for the antiparticle (section 1.3). Curved traces at angular frequencies $\omega > 1$ and $\omega < 1$ represent corresponding kinetic energies (section 1.5).

Meanwhile it will become clear from Figures 1.1, 1.2 and 1.3 that the wave group bounds the carrier wave. This system - whether for a massless photon or for a particle - is a constraint on the relative motions of group and wave. To an observer, the constraint results in quantization upon measurement. A particle at rest has energy $E = m_o c^2$. As kinetic energy is added, E increases as:

$$E = m\,c^2 = \frac{m_0 c^2}{\sqrt{1 - \frac{v_g^2}{c^2}}} \tag{1.5}$$

taking the positive solution for equation 1.1, where m is the relativistic mass. From formulae 1.1-1.5, v_g can be expressed as a function of k:

$$v_g = c\sqrt{\frac{1}{1+\frac{m_0^2c^2}{\hbar^2k^2}}}, \tag{1.6}$$

(Figure 1.3) where a corresponding function can be found for v_p, or more simply $v_p = c^2/v_g$. As expected in relativity $k \to 0 \supset v_g \to 0$ and $k \to \pm\infty \supset v_g \to c$ while the classical case is given approximately by $E \sim p^2c/2m_o + m_oc^2$ at low k, so that the dispersion becomes $\omega \approx k^2c^2/2\hbar m_0$. While $k > 0$ and $k << m_0$, the second derivative of the dispersion has the positive value $c^2/\hbar m_0$. Positive dispersion accompanies right-handedness (RH in appendix A.I) in the wave function of the particle when, for the photon, $m_0 = 0$. Then $v_g = v_p = c$ *in vacuo*, or in a dielectric medium of refractive index n, $v_g = v_p = c/n$ [12]. As in relativity, the direction of propagation is treated separately from the transverse plane: in the propagation direction x, the relativistic energy E_x is given by equation 1.2; but in transverse, the kinetic energy component $E_y \simeq p_y^2/2p_x$, with p_x replacing m_0. The corresponding equation holds for E_z.

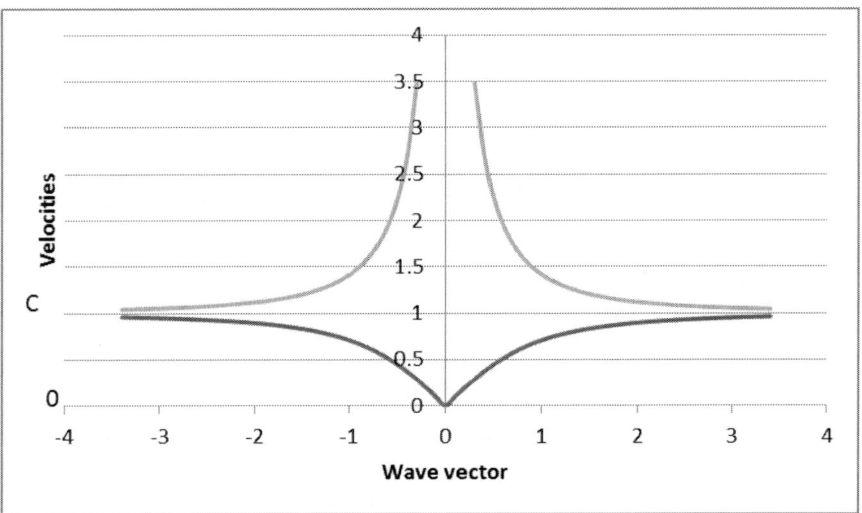

Figure 1.3. Calculated group velocities v_g (dark lines) and phase velocities v_p (grey lines) following equation 1.6 for particles (right, $k > 0$) and antiparticles (left, $k < 0$).

In summary before examining antiparticles, the wave group of the particle expresses dynamics that are, *prima facie*, more or less familiar in classical physics. The phase velocity is measurable, either as the ratio of energy to momentum, or as the inverse of the group velocity. The velocities are also measured or calculated in various ways: as distance travelled per unit time; or as kinetic energy; or momentum; or in magnetism through the Lorentz force; *etc*. Then relativity and quantization require that $mv_g = \hbar k$, where $m = m_0/(1-v_g^2/c^2)^{1/2}$ is the relativistic mass. Moreover, the Pythagorean relationship with mass that is evident in equation 1.2 implies 5 dimensions in space-time [13, 14]: linear propagation; time; 2 transverse planar dimensions; and mass. Omitting the planar dimensions leaves a familiar 3 dimensional system. Electromagnetism is included in this scheme (appendix A.III.2). However, the dispersion dynamics of antiparticles are more complicated.

1.3 Antiparticle Dispersion

Equations 1.1 and 1.2 represent the stable wave packet constrained by relativity, and they have two solutions (caption for Figure 1.2) for a given wave vector k. As with the eigenvalues for Dirac's relativistic, first order, matrix mechanics, one solution is negative and this he gave to the positron. He pronounced this antiparticle to have positive mass and positive kinetic energy. A set of conventions have grown on these assumptions, which are incompatible with equations 1.1-1.6. We therefore have to examine the advantages of our alternative view for the antiparticle.

Sometimes conventions are obviously wrong but innocuous. For example, electric current passes from the positive pole to the negative, even though, in most conductors, the charge carriers are negatively charged electrons that flow in the opposite direction. Ionic conduction is exceptional, as in Li^+ ion batteries, but also in other conductors above normal temperatures. Sometimes holes in p-type semiconductors are said to flow in the direction on the current, but this is an error to be examined in the next chapter. The fault is a natural consequence of a weak convention. However the unphysical singularities that are found in the dispersion of antiparticles (appendix A.II) are of a different order, and that is why the dynamics of antiparticles require more fundamental treatment.

Firstly, if its eigenvalue is negative, the rest mass must be negative, because that is what the energy is. Secondly, if the rest mass is negative and the kinetic energy is positive, then the condition $(|k| \to |m|) \supset ((v_g \to \infty) \cup (v_p \to 0))$ would be unphysical: the conventional supposition, of negative eigenvalues for the antiparticle together with positive kinetic energy is incompatible with ω/k dispersion. Thirdly, $k < 0 \supset \omega < 0$, otherwise the wave function in equation 1.1 would self-annihilate, violating experimental conservation rules. Three prior conventions are therefore inconsistent with equations 1.1-1.6. Moreover, since ω is negative, the eigenvalue has to be consistently negative, which was the initial presumption. This consistency is represented in the third quadrant of Figure 1.2 and in the negative half of Figure 1.3. Notice that v_g and v_p are always positive for both particle and antiparticle. As in bubble chamber and cloud chamber photographs, the direction of propagation is the vector velocity v_g. In this representation the mass and kinetic energy of the antiparticle are both negative, and the second derivative on the dispersion is also negative. Further implications will be discussed in section 1.5 and in the two following chapters. However, it is worth noting that an infinitely extended wave with infinitesimal amplitude will have negligible response to a localized applied force. Measured responses are due in major part to the variations in localized amplitude provided by the envelope packet. This is why, for example, the Lorentz force acts on group velocity; not phase velocity.

Conventionally, the description of matter, including the Lorentz force and magnetism generally, presents the particle as right handed (RH in appendix A.I). This is the positive energy solution for equations 1.1 and 1.2. The negative energy solution, just described, accompanies a set of negative values: E, m, ω and k are all negative, where the vector k opposes the propagation v_g.

1.4. Uncertainty

This is a property of waves. It is often applied as a fundamental axiom in mathematical quantum mechanics, but it is easily and accurately derived physically, in classical wave mechanics. As such, it is an example of how the empirical method in physics produces uncomplicated solutions. Starting with equation 1.1, the probability distribution for finding a particle at position x and time t is given by:

$$P(t,x) = \phi^*(t,x).\phi(t,x) \tag{1.7}$$

By writing $x = 0$ and performing a Fourier transformation with respect to ωt,

$$P(\omega)_{x=0} = \tilde{F}(\phi^*(t).\phi(t))_{x=0}\ , \tag{1.8}$$

the probability distribution for a particle with angular frequency ω is found. The transform of a Gaussian is Gaussian and is given by:

$$P(\omega)_{x=0} = \frac{\sigma}{\omega\sqrt{2}}\exp\left(-\frac{\omega^2\sigma^2}{4\overline{\omega}^2}\right) \tag{1.9}$$

while the probability distribution in time is described by the envelope part of the function in equation 1.1.

$$P(t)_{x=0} = \exp\left(-\frac{t^2\overline{\omega}^2}{\sigma^2}\right) \tag{1.10}$$

Some choice is available in defining Uncertainty, Δt, and $\Delta\omega$. A natural selection is the full width half maximum of the Gaussians in equations 1.9 and 1.10. However to reach out to Heisenberg's lack of definition, we can propose an alternative definition that he might have derived from the ground state of the hydrogen atom. We propose

$$\Delta t = 2\frac{\sigma}{\overline{\omega}} \tag{1.11}$$

i.e. the full width of the distribution at heights of the inverted exponent 1/e. Similarly:

$$\Delta \omega = 4\frac{\overline{\omega}}{\sigma} \tag{1.12}$$

Multiplying the two uncertainties reveals the very significant fact that $\sigma / \overline{\omega}$ cancels. Substituting for ω from the derivative form of Planck's law:

$$\Delta E \cdot \Delta t = 8\hbar \tag{1.13}$$

which is more than an order of magnitude greater than Heisenberg's estimated limit, $\Delta E \cdot \Delta t \geq \hbar / 2$. The difference may not matter for mathematics; but it does for engineering [15, 11]. Further inaccuracies occur in other dimensions, such as the transverse plane. Strictly, equation 1.13 is valid only for the free particle stable wave packet in the direction of propagation; there are additional integration errors in more general cases. It is interesting therefore to compare uncertainties for bound states such as the ground state of the hydrogen atom. The function is given by [16]:

$$\varphi_{1s} = \frac{1}{\sqrt{\pi}}\left(\frac{Z}{a_0}\right)^{3/2} \exp\left(\frac{-Zr}{a_0}\right) \tag{1.14}$$

where Z is the nuclear mass, r radial distance, and a_0 the Bohr radius. This is a more complicated function to transform and define than equation 1.1; but estimates about $\Delta\omega\cdot\Delta t\approx 1$ and $\Delta k\cdot\Delta x\approx 1$ fit the uncertainty relations in a vague way. In fact the stable state is infinite in time and infinitesimal in uncertainty, so the Principle does not apply. However, when you squeeze a wave in time the energy spreads, or squeeze in space to spread momentum. Both actions are typical of waves.

Alternatively, the energy of a photon is derived from Maxwellian theory and is quantized by standing wave conditions or by boundaries, typically at emission or detection:

$$\varepsilon = \int E_f^{\,2} d\tau = \int f(A,\sigma,\omega,k,t,x)d\,\tau = \hbar\omega \tag{1.15}$$

integrated over all space, with E_f the electric field vector. Not only are spatio-temporal frequencies quantized by boundaries, but so also are amplitude and coherence by normalization of the wave function, both in ϕ and in intensity $\phi^*\phi$ for each particle. In the photon, the fields are real though the relative phases of the E_f and magnetic B_f fields are often represented by the real and imaginary parts of a complex function. By contrast, the probability amplitudes of Fermions are properly complex solutions to wave equations. However it is obvious that probability and uncertainty are equally common to both; from them follow a description of mass and acceleration in terms of dispersive wave properties.

1.5. Negative Mass

Figure 1.2 contains an illustration of negative mass, and further detail for the necessary concept is needed. In the figure, negative mass is a property of antimatter. This property has a long history that is discussed by Villata [17] who concluded that "the current formulation of general relativity predicts that, while matter and antimatter are both self-attractive, matter and antimatter repel each other under the assumption that matter is transformed into antimatter by the CPT (charge-conjugation, parity, time) operation." Others have claimed to measure it [18]. Our concern lies within the focus of how to represent the mass in kinetic energy and acceleration, where the medium consists in imposed crystal force fields. We are interested not so much in the mass, as in corresponding dynamics.

First, the group velocity requires definition, not only because it is consequential in the antiparticle, but because it is poorly served in introductory texts. The definition will serve as the prelude to the second derivative, itself vital to the dynamics of mass and force. Consider two plane waves:

$$\phi_1 = A\sin((k+\Delta k)x-(\omega+\Delta\omega)t), \tag{1.16}$$

and

$$\phi_2 = A\sin((k-\Delta k)x-(\omega-\Delta\omega)t). \tag{1.17}$$

with the superposition,

$$\phi_1+\phi 2 = A\sin(kx-\omega t)\cos(\Delta kx-\Delta\omega t). \tag{1.18}$$

This is the equation for a wave with mean wave vector and angular frequency and with a cosine envelope. Fixing x at some point, describes the beats in sound waves used by piano tuners. Waves subject to the dispersion shown in equation 1.3 have phase velocity $v_p = \omega/k$ and group velocity $v_g = c^2/v_p$. A similar cosine envelope is obtained when an exponential function is made to substitute for the sine wave. By further superposition of distributed pairs of waves with appropriate amplitudes, the Gaussian group shown in equation 1.1 is constructed, each pair summing to approximately the same phase and group velocities as the others when $\sigma << k$ because the dispersion is well behaved. The group is therefore stable. This physical description of the group velocity can be derived mathematically from the derivative of the second part of equation 1.4, putting $X = 0$ at the center of the wave group:

$$x.dk + k.dx - t.d\omega - \omega.dt = 0 \tag{1.19}$$

where $dt = 0 = dx$, $d\omega/dk = x/t$ gives the group velocity. Likewise where $t = 0 = x$, $dx/dt = \omega/k$ gives the phase velocity. Furthermore, if s is the location of the group center at $X = 0$, then $d\omega./dk = ds/dt$ and $d^2\omega./dk^2 = d^2s/dt^2$.

Equation 1.2 can be written in terms of relativistic mass $E = mc^2$. The kinetic energy is $E\text{-}m_0c^2$. In Newton's second law, relativistic mass is the ratio of applied force F to acceleration $a = d^2s/dt^2 = dv_g/dt = d^2\omega/dtdk$, where s is the location of the wave group center in equation 1.1 and Figure 1.1. It follows that the second derivative of the dispersion, which is evident in the kinetic energy in Figure 1.1 and plotted in Figure 1.4, is represented by:

$$\frac{d^2\omega}{dk^2} = \frac{dv_g}{dk} = \frac{dv_g}{dp} = \frac{1}{m} = \frac{a}{F} \tag{1.20}$$

This second derivative $d^2\omega/dk^2$ on the first term is positive for the particle where $m > 0$; and negative for the antiparticle where $m < 0$. In the propagation direction, the equation is 1-dimensional, but it can be transcribed easily to 3-dimensions by substituting the first term with $\nabla_k^2\omega$, the dispersive curvature. When we come to apply the curvature in the Hall effect, we will need, for the direction of propagation, the x-component $\nabla_{k_x}^2\omega$.

We arrive at a conclusion that is consistent with the Feynman Stückelberg switching principle: "An antiparticle travelling forward in time does not exist": in dispersion dynamics, its momentum wave vector is antiparallel to its group velocity. This conclusion will be used in the following chapters because of its usefulness in understanding the motion of holes, not only in p-type semiconductors and in HiT_c superconductors, but also in several metals. In all of these cases the negative second derivative is a consequence of crystal fields as we shall see. Other general consequences of negative mass are mentioned in appendix A.III.

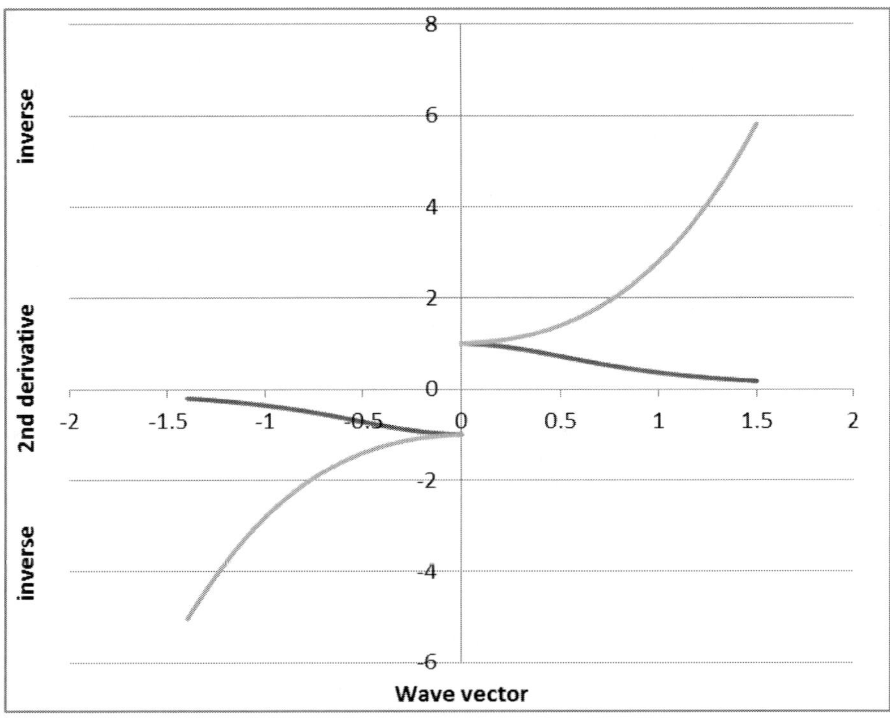

Figure 1.4. (Dark traces at center between ordinates ±1), second derivative $d^2\omega/dk^2$ (equation 1.11) of curved traces in Figure 1.2; and (light traces at outer regions corresponding to ordinates >1 or <1) the inverse of the second derivative. Its curvature is positive in the first quadrant and negative in the third quadrant. Mass is inversely proportional to acceleration in a force field.

Before leaving this chapter, a summary is needed. Some conventions in science are innocent and acceptable. Before the nuclear model of the atom was understood, electric current was set to pass from positive pole to negative. Subsequently it became clear that most currents were due to negatively charged electrons passing in the reverse direction. In particular, when Hall discovered his coefficient, the carrier charges had been thought of as positive. When the charges in his specimens were known to be electrons, his coefficient came out generally negative. Notable exceptions are ionic conductivity, particularly of Li^+ in modern batteries, or more generally in heated fuel cells. For too long, hole conductivity in *p*-type semiconductors was thought to be another exception; but this view is shown here to be confusing and erroneous. Dispersion dynamics are not optional like the flow of electric current. The first objection to conventional theory is the inconsistency illustrated in the unphysical singularities described in appendix A.II and in A.III. Moreover, Heisenberg's uncertainty is notionally vacuous and inaccurate[8]. It is replaced here by empirical wave mechanics. We find that mass, whether in free particles or in crystal fields, may be positive or negative, depending on dispersion relations. Notice that the oscillational frequency that is due to rest mass is constant. The dynamics depend on derivatives in the dispersion while inertia is weighed by mass.

In physics consistency is necessary; completeness, as in Gödel's theorem for any axiomatic system, is unattainable. The following chapters show how dispersion dynamics enable an empirical understanding of HiT_c.

References

[1] Bardeen J., Cooper L. N. and Schrieffer J. R., 1957, *Phys. Rev.* 108 1175.

[2] Bourdillon A. and Tan-Bourdillon N. X., *High Temperature Superconductors: Processing and Science*, Academic Press, N. Y., 1994, ISBN 0-12-117680-0, p. 33.

[3] Batlogg B., Cava R. J., Jayaraman A., van Dover R. B., Kourouklis G. A., Sunshine S., Murphy D. W., Rupp L. W., Chen H. S., White A., Short K. T., Mujsce A. M. and Rietman E. A., 1987, *Phys. Rev. Lett.* 58 2333.

[4] Rickaysen G., *The Theory of Superconductivity*, Interscience, Wiley, New York, 1965.

[5] Schrieffer J. R., *Theory of Superconductivity*, Benjamin, NewYork, 1964.

[6] deGennes P. G., *Superconductivity of Metals and Alloys, Benjamin, New York, 1966.*

[8] Typical of Newtonian corpuscularity, in turn mistaken for quantized wave packets.

[7] Dirac P. A. M., The p*rinciples of quantum mechanics, 4th ed. 1958, Clarendon.*
[8] Harrison W. A., *Solid State Theory*, McGraw-Hill, 1970.
[9] Popper K. R., *The logic of scientific discovery,* 1983 Hutchinson.
[10] Popper K. R., *Quantum theory and the schism in physics,* 1982, Hutchinson.
[11] Bourdillon A. J., *Journal of Modern Physics* 2015 6 2011-2020.
[12] Bourdillon A. J., *Journal of Modern Physics, 2012,* 3 290-296*; doi10.4236/jmp.2012.33041.*
[13] Bourdillon A. J., *Journal of Modern Physics,* 2013, 4 705-711 *(2013),* doi:10.4236/jmp.2013.46097.
[14] Bourdillon A. J., 2014, *Journal of Modern Physics,* 5, [1] 23-28 (2014) doi:10.4236/jmp.2014.51003.
[15] Bourdillon A. J., *Ultramicroscopy,* 2000, 83 261-264 (2000) *invited paper for commemorative issue in honour of Michael Stobbs.*
[16] Pauling L and Bright Wilson E, *Introduction to quantum mechanics,* McGraw-Hill 1935, p. 139.
[17] Villata M. CPT symmetry and antimatter gravity in general relativity, EPL 94 (2) 2011.
[18] Khamehchi M. A., Khalid Hossain, Mossman M. E., Yongping Zhang, Busch Th., McNeil Forbes M., and Engels P., 2017, arXiv:1612.04055v2 [cond-mat.quant-gas].

Chapter 2

The Lorentz Force Does Not Act on Voids

2.1. Positive Hall Coefficients

In the previous chapter, the dynamics of particles in free space were described. Now consider the adaptation of those dynamics to constrained particles, and in particular to particles immersed in crystal fields. In view of later requirements for measurements of those dynamics we have also to find clarity about how the Lorentz force applies. The schematic diagrams in this chapter are chosen to illustrate those applications.

A particle with mass m and charge q, travelling with velocity $\boldsymbol{v}$ inside magnetic induction $\boldsymbol{B}$, experiences an acceleration a that is normal to both its velocity and the field vector. This fact is a result of Lorentz's law which determines the Hall effect, and indeed electrodynamics:

$$ma = q.\boldsymbol{v} \times \boldsymbol{B}. \tag{2.1}$$

When either $q = 0$ or $v_g = v = 0$, the Lorentz force is zero. It acts neither on voids nor on stationery ions. The consequence is important in a proper understanding of materials that display positive Hall coefficients.

Since, in the superconducting state, the electric field, $\boldsymbol{E} = 0$ and the magnetic force field, $\boldsymbol{B} = 0$[9] [1], *i.e.* both zero, we can only measure the charge carriers in the normal state, either at elevated temperatures above the critical temperature, $T > T_c$, or in strong magnetic fields above the critical induction field $\boldsymbol{H} > \boldsymbol{H}_c(T)$. On cooling to T_c, the Fermionic normal state carriers condense to the Bosonic superconducting state. In low temperature superconductors, the carriers are negatively charged and have negative Hall coefficients[10], measured in the normal state; the HiT_c *compounds have positive Hall coefficients*. In this state, the coefficient R_H measures the cross voltage resulting from an electric field E_z that is generated in material with carrier density n, conductivity σ, current density $j_x = \sigma E_x$, made to pass through a magnetic cross field B_y [2]:

$$R_H = \frac{E_z}{j_x B_y} = \frac{1}{nq} \tag{2.2}$$

since $j_x = nqv_x$. This is an application of Lorentz's law on negatively charged electrons where $q = -|e|$ and e is the electronic charge.

In introductory texts, Hall measurements are simply described. Suppose that the charge carriers act as almost free particles. Let the electric field E that is applied across a conductor be vertically upwards $E = E_x$. The carriers are accelerated and are scattered by collisions with the lattice to achieve the mean current density, $j = nv_d q$, where n is the effective carrier density and v_d the drift velocity. Positive charges are forced parallel to E_x; negative charges anti-parallel. When the magnetic force field $B = B_y$ is applied normal to E. The carriers are diverted by the Lorentz force, and so tend to the z direction where charge buildup causes a cross field $qE_z = qv_x B_y$. In the steady state, $j_z = 0$. Then the Hall coefficient is defined as in equation 2.2. R_H is negative for electron charge carriers and positive for ionic conduction. It is often supposed that holes in the conduction bands of p-doped semiconductors are positive carriers [3], but general objections

---

[9] This is the Meissner-Öchsenfeld effect. There is, however, local penetration in type II superconductors including the high temperature superconductors.

[10] When Hall discovered his effect, electric currents in metals were supposed to carry positive charges.

can be raised to the over-simplified explanation. In order to interpret the published Hall effect data on HiT_c ceramics, we will consider one show stopper first, and then two criticisms. The solution follows the dispersion dynamics of localized particles and antiparticles that were described in the previous chapter. The result applies in the same way to positive Hall coefficients measured in *p*-type semiconductors and in several metals, including *Al*.

The principle objection to be considered is that the holes that will be described in chapter 3, are not particles: the *Lorentz force does not act on voids.* The hole 'carrier' was supposed to be a vacancy in an otherwise filled band of states. The states are derived from fine atomic-like states that, in the crystal, are dispersed in energy by the periodic lattice potential. The hole vacancy in the valence band was supposed to act like a particle. How?

A similar argument applies to immobile ions which are likewise inactive in determining the Hall coefficient. Though they resistively scatter electrons in the HiT_c compounds, they are fixed in the crystal lattice: the *fixed* ions do not transport electric current at either normal or low temperatures in any of the materials being considered, namely superconductors, *p*-type semiconductors or regular metals (cf. mobile Li^+ ions and other ionic conductors). Since negative coefficients result from magnetic forces on conducting electrons, how are positive coefficients produced in HiT_c compounds? Before finding the solution, it is helpful to refer to two further objections.

Secondly, several metals, including one of the most conductive, namely *Al*, have positive Hall coefficients (Table 2.1 [2]), even though there is no doubt that the carriers are negatively charged electrons. It should be assumed that the mechanism that drives the positive coefficient in *Al* and several other metals is the same as that operating in *p*-type semiconductors and in HiT_c superconductors. The next section will show what that mechanism is. By contrast, positive R_H values are expected from Li^+ ions and H^+ ions when they sometimes conduct. They do this even at normal temperature, owing to their comparatively small size and large ionization energies. Other ionic conductors transport current at elevated temperatures.

The third criticism derives from complications that arise in strong magnetic fields: the free paths between scattering collisions cease to be approximately linear and tend toward circular resonance. This affects magneto-resistance and stimulates important information on Fermi surfaces in metals (*e.g* [3]). The foregoing analysis is done in weak field approximation, but the general information given by the action of strong magnetic fields on the electron energy bands in many solids provides background for the general solution.

Table 2.1. Low field experimental Hall coefficients measured conventionally at room temperature [4] or by helical wave at 4 K [5], and calculated carrier densities [6]

Metal	Method	Experimental R_H	Calculated $1/nec$
	Helical wave at 4K (H), or Conventional at room room temperature (C).	$\times10^{-24}$ CGS units	$\times10^{-24}$ CGS units
Li	C	-1.89	-1.48
Na	H	-2.619	-2.603
	C	-2.3	
K	H	-4.946	-4.944
	C	-4.7	
Rb	C	-5.6	-6.04
Cu	C	-0.6	-0.82
Ag	C	-1.	-1.19
Au	C	-0.8	-1.18
Be	C	+2.7	
Mg	C	-0.92	
Al	H	+1.136	+1.135
In	H	+1.774	+1.780
As	C	-50.	
Sb	C	-22.	
Bi	C	-6000.	

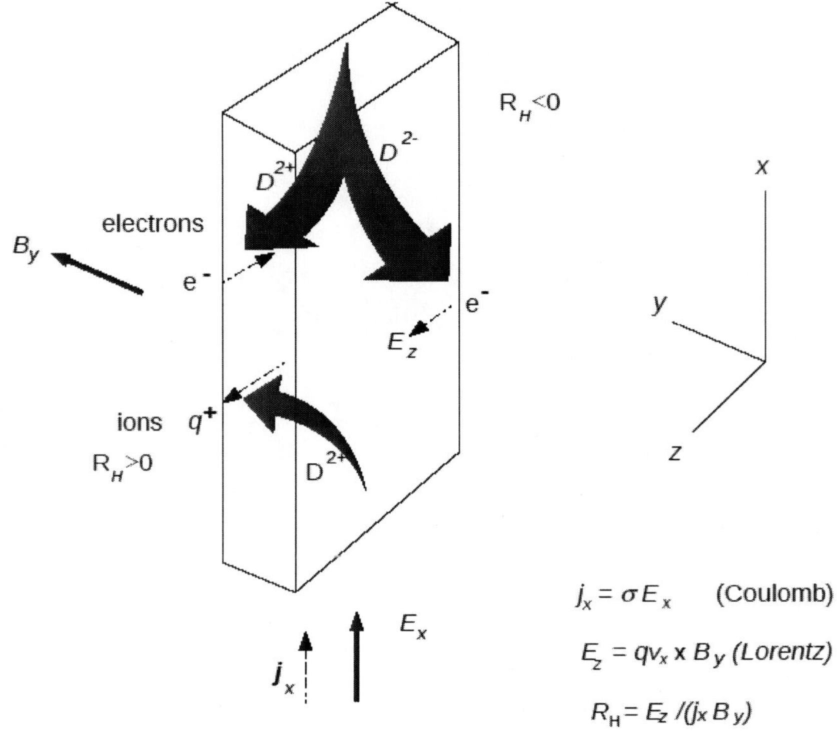

Figure 2.1. Transient currents in three types of specimen with corresponding Hall fields E_z (dashed arrows) or steady state voltages. Electrons are accelerated downwards by the upward electric field E_x; positive ions accelerate upwards with positive R_H. Electrons in negative dispersive curvature $D^{2-} = \nabla_x^2 \omega < 0$ have $R_H > 0$ like ions; when $D^{2+} = \nabla_x^2 \omega > 0$, their $R_H < 0$, as is more common in metals.

These problems own a common solution. Remember that Equation 2.2 demonstrates a physical property that connects the measurables E_x, B_y and E_z; but n and q that are each inversely proportional to R_H, are soft numbers that are not independently measured. Of these, n is a number counter and can only be positive if it has any meaning, while q must be the electronic charge -|e| because the Lorentz force can act on neither voids nor immobile ions. It is inescapable that R_H can become positive only by a change of effective mass that causes v_x to change sign. Equation 1.20 shows how this is done. In electrodynamics generally, e/m is invariant, so that a change of sign in one is effectively equivalent to a change of sign in the other. Instead of q changing sign for the hole, the Lorentz force is preserved by:

$$R_H = -\frac{m_e}{m_{eff}}\frac{1}{ne} \tag{2.3}$$

where m_e is the rest mass of the electron and m_{eff} follows equation 1.20. The charge is always negative and the effective mass may be +ve or –ve, depending on the dispersive second derivative $\nabla_k^2\omega$. For the simple case of perfectly free electrons, $m_e = m_{eff}$.

The Hall coefficient is therefore proportional to the mean curvature near the Fermi surface in the Brillouin zone, of the dispersion $\nabla_k^2\omega$. Notice that, since the Lorentz force is proportional to velocities generated by electric fields, the electric currents depend on q in first power, whereas the corresponding magnetic deflection is proportional to q^2. However the hole concept is only notional because the Lorentz force does not act on voids; the 'hole' that is often used as a supposed explanation, actually is concomitant with the second derivative in dispersion $\nabla_k^2\omega$. When its curvature is negative, the coefficient is positive, and *vice versa*. How the oversimplified 'hole' description gives a result similar to the proper physical description is illustrated in Figure 2.1. After reconsideration of conductivity in metals and semiconductors, the clarification will be needed for the HiT_c ceramics.

2.2. Typical Metals

Most metals have negative Hall coefficients. This is because conduction is by negatively charged electrons, and especially true when they are nearly free electrons. A few metals (Table 1.1) have positive coefficients, and this is due to periodic crystal fields to be described below.

In particular, *Cu* is typical and its electron energy band structure is well studied. It has further usefulness here for comparison with ionic copper that is critical for superconductivity in the cuprate HiT_c ceramics.

Moreover, it has the same face centered cubic (FCC) crystal structure as *Al,* which has the alternative positive R_H. Crystals consist in periodic arrays of atoms, whose core electron wavefunctions are calculated in the normal way for atoms, using Schrödinger's equation. The motion of the outer electrons is largely determined by the periodic electric and magnetic fields of the crystal. These cause corresponding electron energy levels to spread into bands of states that reach across the crystal, and that are wavelength dependent. The periodic crystal potential forces the outer electrons into Bloch wave functions with periodicity l:

$$\psi_k(r+l) = e^{ik.l}\,(\psi_{k(}r)) \quad (3)$$

where ψ_k repeats from zone to zone. Various computational devices are used to solve for these states. Typically the atomic cores are represented by muffin tins and potentials are constructed between them. These potentials act on valence or conduction electron bands. Since the electrons themselves affect the potentials, calculations must be self-consistent and normally include exchange interactions. Energy bands are the calculated solutions for the wave equations in vector space.

Methods of solution are adapted for various properties investigated [3]. Generally, symmetries in the problem allow reductions in scale to a reduced Brillouin zone. This zone corresponds in *k*-space to the unit cell in the real space of the crystal, where the energy bands cross Brillouin zone walls. For the FCC structure, the zone is labelled in Figure 2.2 as an illustration of points of symmetry and lines of symmetry. Calculations proceed from points of high symmetry to lines of lower symmetry and on to connecting surfaces. Essentials of the calculations are economically represented by the line Figures.

As a simple example, Figure 2.3 is a schematic electronic band structure for the outer shells of *Cu,* with atomic configuration [Ar]$4s3d^{10.}$ The scheme is derived from various detailed band structure calculations [7], and is used to illustrate important features. Beyond the core are two types of electronic band. The filled 3d shell forms a set of states in a comparatively tight band at the bottom of the valence band. Above this band lie the 4s type of Bloch state. Having a comparatively large mean radius, this state is 'nearly free' and has approximately parabolic dispersion based on the zone origin at the Γ point. The dispersion of the conduction electron band is therefore nearly free, like that of free electrons at low *k*, on the positive side of the ordinate axis in Figure 1.2. The bands conduct electrically where they intersect the Fermi level at energy E_f, partly due to thermal activation that increases the population of conduction electrons. In three dimensions, the Fermi surface is, in first order, roughly spherical. On the parabolic conduction bands, the dispersive curvature is positive, $\nabla_k^2 \omega > 0$, so the Hall coefficient is negative.

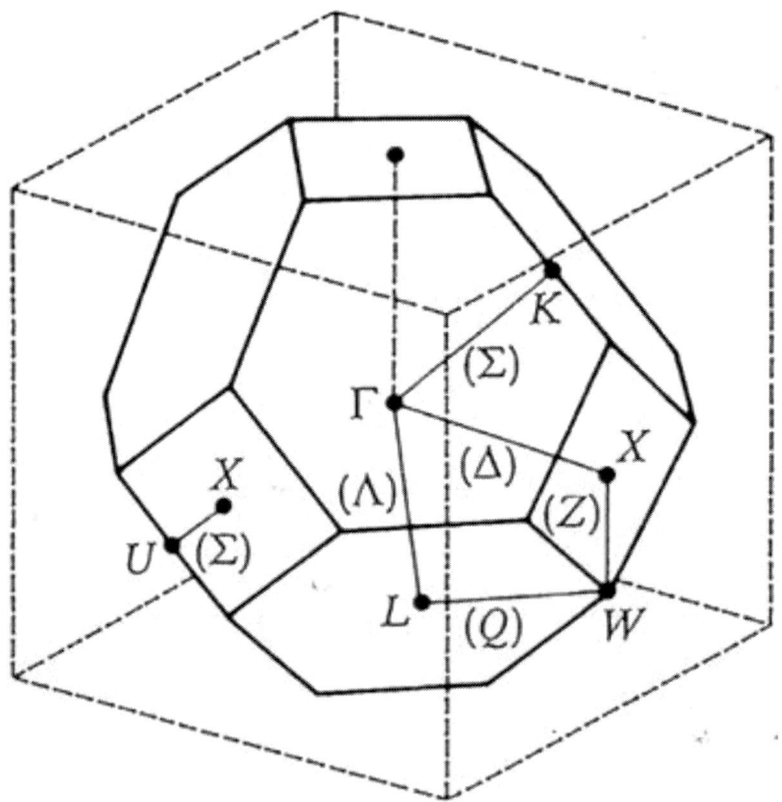

Figure 2.2. Brillouin zone for the FCC structure in *Cu* and *Al*, showing points of symmetry and lines of symmetry. The same zone in wave vector space applies to the cubic diamond structure in *Si*.

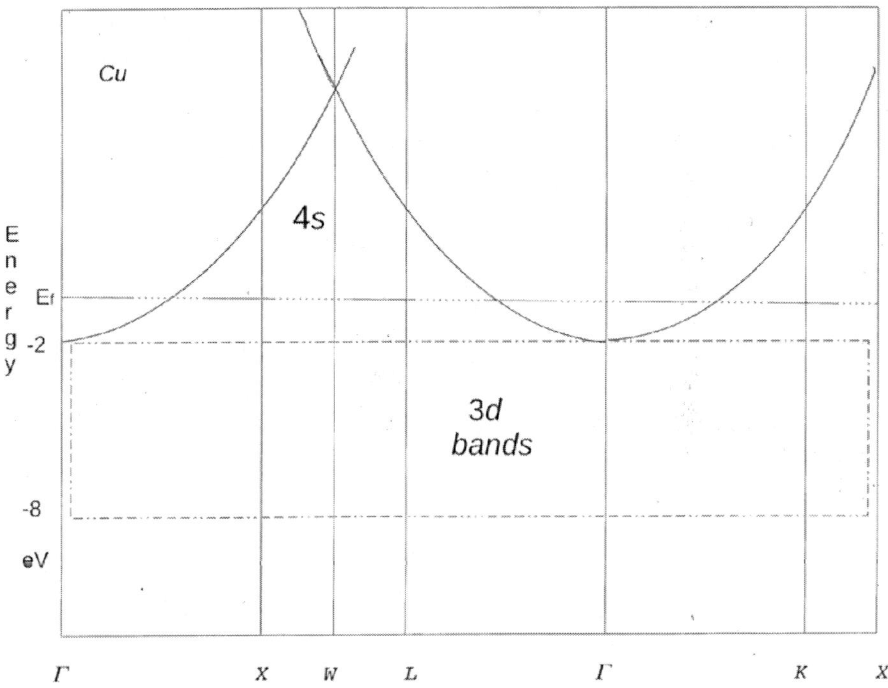

Figure 2.3. Schematic electronic energy band structure of pure *Cu*, showing the nearly free, parabolic, 4*s* band loosely bound on top of the filled $3d^{10}$ shell. The Fermi level (dash-dot horizontal line) passes through the 4s band where the dispersive curvature (measured from the Γ point) is positive.

The many difficulties that exist in calculating reliable band structures and of interpreting conductivity from them, is partial justification for employing the schematics to illustrate the general features of the Hall effect. Happily, the following example of *p-type* semiconductors is relatively simple, but the conductivity is again due to negatively charged electrons, as Hall measurements with the Lorentz force very well show.

2.3. P-Type Semiconductors

In metals, electronic states continue unbroken from valence to conduction bands. In semiconductors the bands are separated by a gap of

0.5-2.0 eV [8]. On the simplest Kronig-Penny model for the crystal, the gap is bounded by valence periodic waves in phase with lattice potentials, while the conducting upper boundary is in antiphase. In pure, intrinsic semiconductors, the Fermi level lies near the gap center, and electronic conduction is activated by thermal electrons excited from the valence to conduction bands. Then, both free electrons in the conduction band and freed holes in the valence band contribute to the conduction. Notice that *Si* has an indirect band gap (Figure 2.4): the minimum in the conduction band occurs at a wave vector that is different from the vector at the peak of the valence band at Γ. Excitation of the thermal electrons therefore requires phonon assistance and is slower than in direct gap semiconductors. Notice also that while the curvature at the bottom of the conduction band is positive $\nabla_k^2 \omega > 0$; the corresponding curvature at the peak of the valence band is opposite $\nabla_k^2 \omega < 0$. The free electrons in opposing bands (Figure 2.1), when under the Lorentz force therefore move in opposite directions. The conduction electrons in the former case make a negative contribution to the Hall coefficient R_H; while the latter contribute positively.

The resultant depends on the relative mobilities of the two types of carrier. P-type semiconductors are doped with acceptor atoms, having lower valence than the matrix semiconductor, *B* in *Si* for example. The dopants occupy sites in the band gap close to the valence band (the crosses in Figure 2.4). Electrons vacate these sites at very low temperatures, leaving them positively charged. However, the sites are typically thermally activated, leaving vacancies or 'holes' in the valence band. Thermal activation neutralizes some of the dopant charge leaving mobile electronic states on the otherwise filled valence band. The dopant sites are sparse, but the holes are extended and mobile. They move with positive Hall coefficient R_H because the curvature of the band near the dopant sites is negative, $\nabla_k^2 \omega < 0$. The holes have an effective positive charge, but the motion that is detected by the Hall apparatus belongs to electrons with negative effective mass.

N-type semiconductors are doped with donor atoms replacing acceptor atoms. Phosphorus is a common donor. Its energy state occurs close to the conduction band, the Fermi level moves up to that level and thermal activation, into the positively curved conduction band, occurs in the normal way. The corresponding R_H is negative and unproblematic.

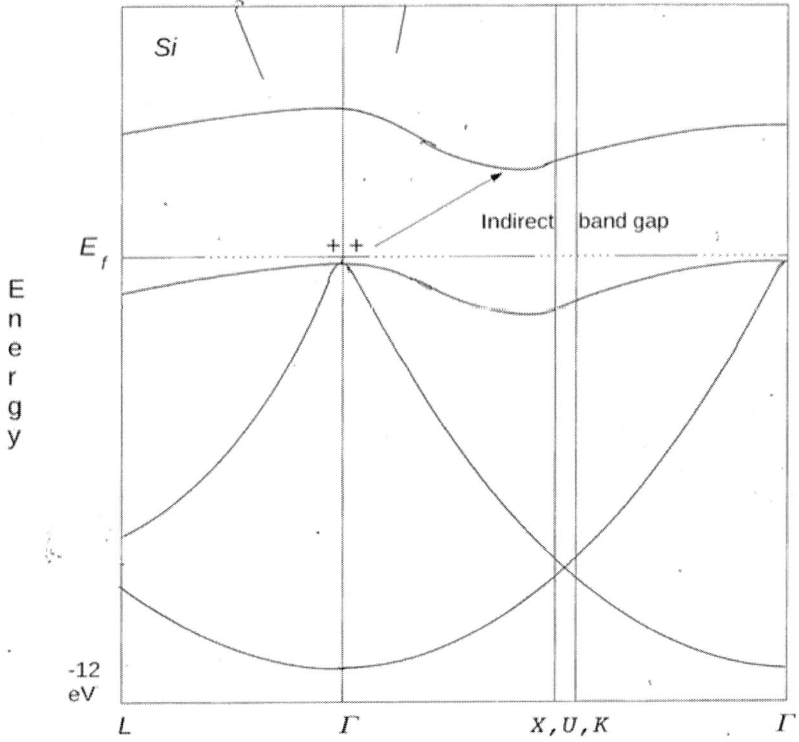

Figure2.4. Schematic electronic energy band structure for p-doped *Si,* with the Fermi level (dash dot) lying atop the valence band. Holes are left in the valence band by electrons, thermally excited into the impurity states (crosses). Conduction is due to passage of electrons, with negative second derivative in dispersion between the holes.

2.4. Metals with Positive Hall Coefficients

Of the metals that are listed in Table I, having positive Hall coefficients, *Al* is a good example both because it has the highest normal conductivity, weight for weight, of any metal; but also because it has a well-studied electronic energy band structure that can be inter-related with data.

Atomic *Al* has the electronic configuration [*Ne*]$3s^2 3p$, *i.e.* one *p* shell electron less than has the *Si* discussed in the previous section. By comparison with *Cu*, the configuration is significant. The band derived from the 3*s* electrons are again parabolic and free electron like, while the *p*-like bands are flatter. However, contrary to the case in *Cu*, the *s*-like bands lie below the *p*-like bands in *Al.* The bands hybridize and form gaps, like those in *Si,* but this time not all bands are discontinuous (Figure 2.5). Some are calculated to pass seamlessly from valence to conduction through the Fermi level [9].

As before, band dispersion occurs with both positive and negative curvatures, and when the latter type lies close to the Fermi level, it determines the sign of the Hall coefficient. R_H is positive when the curvature is predominantly negative, and *vice versa* (*cf.* [10]). Holes or not, the charge carriers are electrons.

Free electron like 3*s* bands are evident in the figure but negative dispersion occurs about the *W* and *K* symmetry points close to the Fermi level. These contrast with the slight positive dispersion in other conduction bands. The experimental, positive Hall coefficients for *Al* point to an uncertainty in the estimate for the Fermi level calculated by Segal. When his estimate is reduced by 0.6 eV, to the level shown in the figure by the dashed line, dispersion in the lower conduction band overlaps the upper valence band. The dominant negative second derivatives in dispersion result in the positive Hall coefficient that is consistent with experimental values for R_H in Table 2.1 and equation 1.20.

The free particle physics already described are sufficiently fundamental that they are capable of throwing doubt on details of band structure calculations in solids. With the strength of the equations, the calculations can be modified to make the calculations consistent with a large body of data including the Hall coefficients. In chapter 3, this understanding will be applied to molecular band structures in HiT_c compounds. Happily, electron energy band structures describe precisely the information that governs the dispersion dynamics of the charged carrier particles.

2.5. Energy Bands in Ceramic HiTc

Band structures for the ceramic superconducting compounds are more complicated than those of elementary metals. The number of atoms in the unit cell is both greater and more varied. The greatest variation is between the cations and anions, but even the cations lie on differentiated states: notably the *Cu* ions on either the chain reservoirs or superconducting planes; but also the various oxygen ions. The valence bands, that are critical to electronic behavior, are made up of states that are based on various anionic oxygen sites.

In particular, the *Cu* ions have a different configuration from elemental *Cu*. The ion is polyvalent with a tendency towards divalency: $[Ar]d^9$, with the outer 4*s* electrons ionized. The 3*d* shell splits predominantly into valence bands, with upper *d*-like states in the conduction band. The valence band also contains outer ionic electrons on oxygen, and their states vary in energy with various Madelung potentials so that they vary with oxygen loading not only in quantity but also in quality. In the next chapter these ambiguities will enter further discussion in the measurements that are made on the superconducting materials. Of particular interest is the curvature of the valence band that is due to either the holey *Cu* cations or *O* anions. If, with all the variables and approximations that must be used, a reliable band structure is incalculable, molecular orbitals may aid comprehensible explanation.

There is however an indication in Figure 2.3 and in more detailed band structures [7] that, with the *Cu*4*s* level raised to the conduction band, negative dispersion should occur at some point equivalent to *X* in the Brillouin zone, and this may explain the conductivity, the Hall effect, and by extension superconductivity in the planar cuprates. In the next chapter, measurements will be described that further define the electronic states.

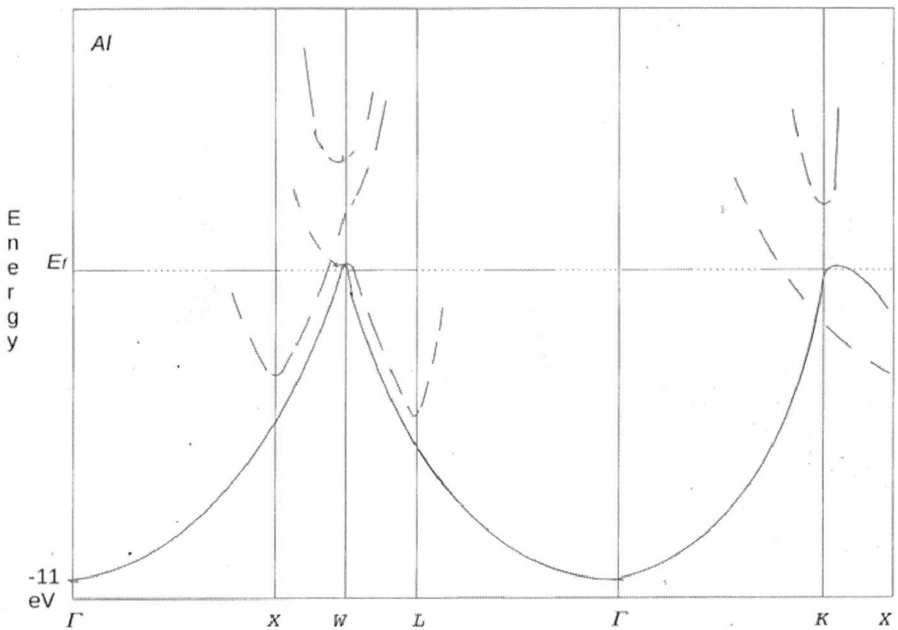

Figure2.5. Schematic electronic energy band structure for *Al* metal with its Fermi level adjusted to lie where the conduction occurs by mobile electrons with predominantly negative second derivative. This is consistent with Hall measurements. Notice the *Al* atom has one electron more than *Si*, and that the parabolic *s*-bands (full lines) lie generally *below* the 2*p*-bands (dashed); whereas they lie *above* the 3*d*-bands in *Cu* (Figure 2.3). The last has the regular $R_H < 0$ and positive dispersive curvature.

REFERENCES

[1] [Bourdillon A. and Tan-Bourdillon N. X., *High Temperature Superconductors: Processing and Science*, Academic Press, N. Y., 1994, ISBN 0-12-117680-0, p. 10.

[2] Hurd C. M., 1972, *The Hall effect in metals and alloys,* Plenum.

[3] Harrison W. A., *Solid State Theory*, McGraw-Hill, 1970 p. 260.

[4] Landolt-Bornstein tables Vol II. 6 p. 161 (1959).

[5] Goodman JM, 1968, *Phys. Rev.* 171 641.

[6] Kittel C, *Solid State Physics,* 5th ed., (1976) p. 176.

[7] Segall B, 1962 *Phys. Rev.* 125 109

[8] Herman F., Kortum R. L., Kuglin C. D. and Shay J. L., in *II-IV semiconducting compounds 1967 International conference,* ed Thomas D. G., Benjamin 1967 New York.

[9] Segall B., 1961, *Phys. Rev. 124* 1797.

[10] LeBoeuf D., Doiron-Leyraud N., Levallois J., Daou R., Bonnemaison J. B., Hussey N. E., Balicas L., Ramshaw B. J., Liang R., Bonn D. A., Hardy W. N., Adachi S., Proust C. and Taillefer L. 2007, *Nature* 450 533-536.

Chapter 3

CHEMICAL HOLES IN CERAMIC SUPERCONDUCTORS

"The fact that the carriers are positively charged is the most important feature which distinguishes high temperature superconductivity from low temperature superconductivity [1]."

3.1. INTRODUCTION

This view has long been acknowledged *e.g* [2], which does not prevent it being wrong. The question outstanding is, "What is the superconducting state?" The associated 'holes' are unscreened nuclear charges on immobile atoms. They are therefore not influenced by the Lorentz force. This acts neither on voids nor on immobile charges; it acts on moving charged particles. The necessity implies that the Hall coefficient R_H should, in the ceramic HiT_c compounds, be negative always. Measurements of positive $R_{H,}$ wherever they occur: in metals, in semiconductors, or in superconductors, require consistent explanation. We find it in dispersion dynamics derived from the stable wave packet [3, 4]. Its logic follows:

The concept of the measured 'hole' (h^{phys}) for the positive Hall coefficient needs modification. The positive coefficient is a consequence of dispersion dynamics: it is due to negative second derivatives in the energy bands of electronic states that are not completely filled. Similar dispersion occurs in the dynamics of free antiparticles along with some metals, p-doped semiconductors, and HiT_c superconductors.

A weakly related definition for the chemical hole (h^{chem}) is conceptually independent, *i.e.* unbalanced local charge in a unit cell, with excess cationic valence over anionic valence. The concept is valid in ionic materials.

In their normal states ($T > T_c$), HiT_c compounds transport normal current with positive R_H. By contrast, low critical temperature superconductors (LoT_c) transport normal current as free electrons with negative R_H.

The chemical hole, h^{chem}, is critical to processing of HiT_c compounds. As such it has the same importance as the lattice distortion in the BCS theory of LoT_c. It follows that the holes are available, in HiT_c, to bind Bosonic electron pairs.

This superconductor pair bonding is principally electrostatic as in compound formation. (A supplementary magnetic component is possible, as with antiferromagnetism, which typically occurs in compounds over-doped beyond the superconducting phase [5, 6, 7, 8]. The vast majority of antiferromagnets do not superconduct).

The holes are products of more than one mechanism. In one of these, anionic overloading of a planar reservoir in $YBa_2Cu_3O_{7-x}$ (YBCO) draws away electrons from holes in adjacent superconducting planes. By another mechanism, mixing with low valence dopants, before solid state reaction in $La_{2-x}Ba_xCuO_{4-y};$ (LBCO; $La_{2-x}Sr_xCuO_{4-y}$ is a similar compound) creates holes more directly. This resembles acceptor doping in semiconductors. Sometimes the holes are produced by planar crystal structures without cationic doping, as in $Bi_2Sr_2Ca_nCu_{n+1}O_{6+2n}$ (BSCCO) and $TlBa_2Ca_{n-1}Cu_nO_{2n+2}$ (TBCCO); though anionic concentrations are influenced by ambient oxygen pressure during baking.

The theory for superconductivity described by Bardeen Cooper and Schrieffer (BCS) depended on the isotope effect: the ratio of critical temperature to isotopic mass, $T_c/M_I^{1/2}$ = constant [9]. Their theory defined the principal features of low temperature superconductivity in metals, and demonstrated the force that binds Cooper pairs, namely a mutual attraction, enabled by a lattice distortion. On this attraction depend the second order phase transition; electronic specific heat variations with temperature; the Meissner effect ($B = 0$); and infinite conductivity ($E = 0$). In ceramic superconductors however, the isotope effect is observed only weakly [1, 10]; instead we have to show how the carrier charges generate and propagate. More recently, new systems have added to our knowledge: the iron pnictides for example; but their T_cs are all lower than the highest in the perovskite and pseudo-perovskite cuprates that were studied in the eventful first eight years of HiT_c research. The subsequent added knowledge has not been revolutionary, and attempts to raise T_c have not progressed. However we now have a clearer demonstration of carrier differences that occur in either LoT_c metals and alloys or HiT_c ceramics. We can therefore better interpret the measurements that have been recorded. The superconductive mechanism that is consistent with the prime experimental features is now more obvious than the quotation atop this chapter describes.

The charge carriers in HiT_c superconductors are called 'holes,' like those in *p*-type semiconductors, where the carriers at normal temperatures are also supposed to be positively charged (by simplification; but see below). The 'holes' contain two meanings: one defined by chemical composition; the other by physical measurement. Without extrinsic carriers, stoichiometric oxide superconductors - where the cationic and anionic valencies balance - [1 p. 2] typically, are insulators. However the doping, that is often employed in the ceramic superconductor processing, results in various metallic, superconductive and antiferromagnetic states [5, 10]. A consistent evaluation is needed for the charge transport. The investigation has two parts: the first describes chemical composition; the second physical measurement. There is obvious provision for superconducting pair bonding and the corresponding superconductive energy gap: the Coulomb force, already responsible in the compound, layered, ionic crystals for chemical hole formation, is readily available to bond pairs of superconducting charges.

3.2. DOPING

It is an empirical fact that not all HiT_c superconductors are doped (see below); but all contain holes. The doping may be *n*-type, donating *free electrons* to the conduction band; or *p*-type, donating minority *holes* to the valence band. Hole donation occurs in two ways: cationic and anionic. Substitution of divalent cations for trivalent, causes holes to be formed in the valence band. Conversely, electron donation occurs when cations of higher valence substitute for lower valence. As an anionic dopant, oxygen loading donates holes to the valence band of a superconducting layer by pulling electrons to an adjacent charge reservoir layer; contrary depletion donates electrons to the conduction band. Being structurally unstable, the 'chains' are charge reservoirs for the comparatively stable 'planes'.

In semiconductor physics, the dopants that cause hole type carriers are called *p*-type, where minority vacancies in a valence band are in thermal equilibrium with nearby *p*-type impurity states having inferior valence. Free electron carriers in the conduction band of semiconductors are in thermal equilibrium with *n*-type impurities, having superior valence to a *Si* matrix, for example. As we show below, the 'positive' *p*-type charge carriers, that operate in the Hall effect, are actually electrons.

The best and most studied example of oxygen loading, $YBa_2Cu_3O_{7-x}$ (YBCO), is historically the first of the HiTc superconductors to transition above the boiling point of nitrogen. By baking at controlled pressures and temperatures in oxygen or in vacuum, x may range from 0 to 1 (Figure 3.1). In the former case T_c = 93 K falling as the oxygen content decreases [1 p. 45]. In this system, the superconductivity occurs only when holes are present. In the range $0 < x < 0.5$, the excess of oxygen ions forms holes h^{chem} in the electronic band structure. When $x = 0.5$, ionic charges balance at 13 per unit cell, and the compound does not superconduct. When $0.5 < x < 1$, the depletion in oxygen results in *n*-type charge carriers, at least from a chemical point of view. In section 3.3 we will discuss their general physical measurement.

To view these changes structurally, consider Figure 3.2 and notice that the HiT_c compounds are layered. The figure may be more or less well known, but it is needed here to illustrate charge reservoir and hole donation in the unit cell of the layered compound. These are important features in the hole exciton that will be discussed later. In the superconducting state with $x = 0$, the CuO_2 atoms are sandwiched in two *planes*, each separated on the one side by interspersed Y^{3+}; and on the other by insulating *BaO* planes. Between the *BaO* planes of neighboring cells lie *CuO chains*. In the stoichiometric state, with $x = 0.5$ after baking in a low oxygen ambience, O^{2-} ions become depleted from the chains, making them effectively $^1/_2[Cu_2O]$ chains. This accompanies a phase transition from orthorhombic to tetragonal. The process is reversible, so that by baking in an atmosphere of oxygen back to $x = 0$, dissociated O^{2-} ions are absorbed into the chains and draw electrons from the CuO_2-Y-CuO_2 planar complex where holes are created. This fact is illustrated by calculations on Madelung potentials [11, 12, 13][11]. In these calculations, charges must balance across a unit cell as indicated at the right of the figure. However, in practice the fractional charges shown on the CuO_2 planes are in principle integral and are balanced by the h^+ hole in the superconducting layer. Here the charges balance when $Y^{3+} + h^+$ provide 4 positive charges that balance the 4 negative charges on the two $[CuO_2]^{2-}$ planes. *The hole is an excess of nuclear charge at the valence band.* The chief evidence is of course the positive Hall coefficients measured at $T > T_c$, due to normal currents. By vacuum baking all the way to x = 1, a reversal depletes the chains of all *O* and releases electrons into the conduction band of the system. The chains therefore serve as charge reservoirs that create lattice holes for superconductivity, especially when x ~ 0. When $T > T_c$, the holes result in a positive Hall coefficient (as below); but on cooling to $T < T_c$, they are available to bind Cooper pairs for superconducting transmission.

---

11 In ref. [13], table I, columns 3 and 5, the consistency of the numbers demonstrate positive 'hole' charges (electron energy band vacancies) on the CuO_2 planes. For a fuller interpretation, the Madelung potentials are supplemented by work functions and electron affinities as described there and elsewhere [13,14].

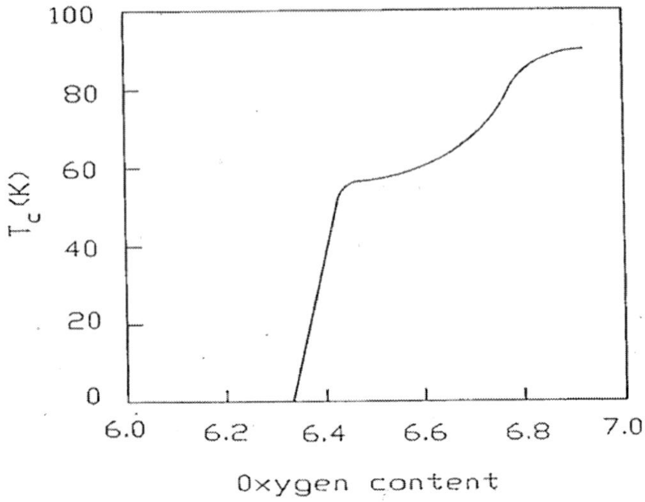

Figure 3.1. Dependence of T_c on oxygen content in YBCO [1].

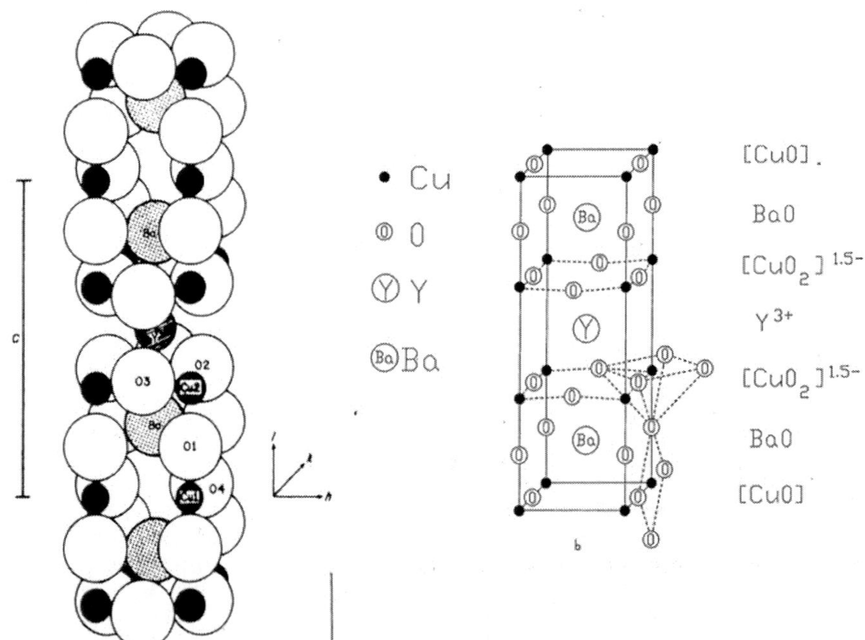

Figure 3.2. Crystal structure of $YBa_2Cu_3O_7$ [1 p. 43] showing atomic packing (left) and coordination (right), together with the planar charges on non-equivalent sites in the layered structure. The fractional charges on the CuO_2 planes include holes induced by the CuO chains as implied by respective Madelung potentials [1, 11, 12].

YBCO is an example of a ceramic superconductor in which the doping is reversible. More generally, doping that occurs during solid state reaction is not easily reversible, but over a restricted range of temperature the anionic composition can often be adjusted. An example is the first discovered HiT_c superconductor, Ba doped La_2CuO_8, or LBCO. This compound has been compared elsewhere [1] with BSCCO and with TBCCO neither of which contain cationic doping: here the holes are a product of the layered crystal structure. Of these two, the former is illustrated in Figure 3.3, showing three forms with $n = 0,1$ or 2. They have T_c = 10K, 80K and 110K respectively. The charge structure in reservoir and hole layers is indicated at the right for the case n=2. For Madelung potential calculations, positive and negative ionic charges must balance across the unit cell as shown. However, electrons attracted to the $[BiO]^+$ layers leave two holes in the cuprate region: the 6 ionic charges on $[CuO_2]^-$ are balanced by $2Ca^{2+} + 2h^+$. The Fermi level is adjusted, and this determines the energy of the hole states in the conduction band. (It is in principle possible to measure the Fermi level by ionization of core excitons in soft X-ray spectroscopy [11, 14] but the same difficulties arise as in photoelectron spectroscopy where various atomic sites, especially of Cu, give mixed signals).

In contrast to the minor structural changes in oxygen loaded YBCO where the hole layer, within the CuO_2 planes, is relatively thin; BSCCO demonstrates the importance of structure in the hole layers, since T_c depends strongly on how thick these layers are, *i.e.* increasing with *n*. The similar TBCCO system, $TlBa_2Ca_{n-1}Cu_nO_{2n+2}$, likewise shows an increase in T_c from 10K to 78, 120 and 121K, when $n = 1, 2, 3$ or 4 respectively; but falls back to 105K when $n = 5$ [1, p. 48]. So in this system there is an optimum thickness for the hole layer complex. The lower critical temperature in YBCO is typical for a cuprate HiT_c structure with n = 2. The composition of reservoir layers has significant but less dramatic effect on the magnitudes of T_c. The same trends are seen in members of the $A_2B_2Ca_nCu_{n+1}O_{6+2n}$ system where *A* and *B* signify a range of cations [1]. The detailed dependence of the superconducting energy gap ε_g on either the oxygen loading or on the structure of the hole layer complex are still unclear; but there can be little doubt that the distribution of charge across the unit cell is fundamental in the formation of superconducting Cooper pairs. The distribution is subsidiary to the prior crystal structure of the multi-layered cuprates.

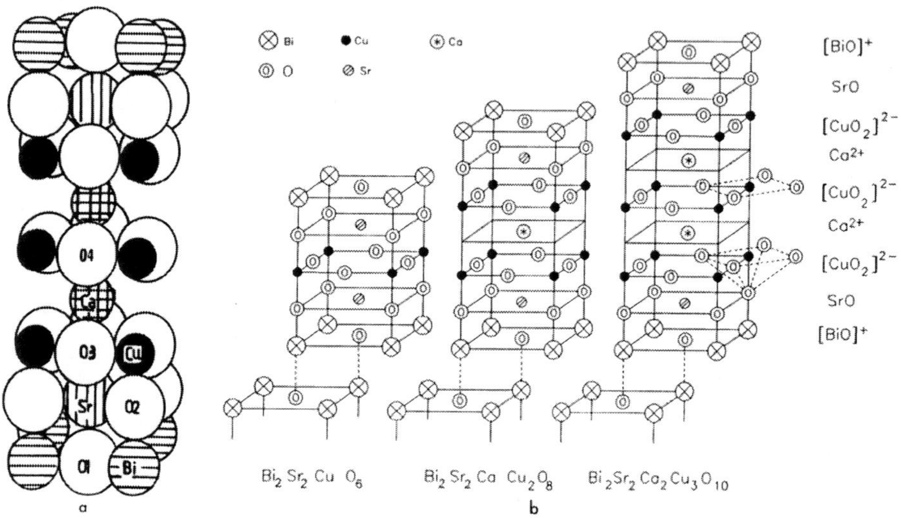

Figure 3.3. a) Atomic packing of BSCCO. b) crystal structures of $Bi_2Sr_2Ca_nCu_{n+1}O_{6+2n}$, with n = 0,1 and 2 showing coordination. At right, planar charges show non-equivalent sites with n = 2.

In BSCCO, the holes were formed without doping, though ambient oxygen pressure is controlled in baking. Maple [5] has described two systems with similar crystal structures: $Nd_{2-x}Ce_xCuO_{4-y}$ is electron doped, since *Ce* has higher valence than *Nd* (like *n*-type *Si* doped with *P*); while $La_{2-x}Sr_xCuO_{4-y}$ is hole doped since *Sr* has lower valence than *La* (like *p*-type *Si* doped with *B*). *The former is a free electron, n-type superconductor; the latter a hole, p-type.* At optimum doping, $Nd_{2-x}Ce_xCuO_{4-y}$ *is a low temperature superconductor with* T_c = *~25 K,* at x = *0.15,* and $R_H < 0$; while $La_{2-x}Sr_xCuO_{4-y}$ *is a high temperature superconductor with* T_c = *~40 K,* at x = *0.18,* with $R_H > 0$. Both systems exhibit antiferromagnetic phases at high x, while the hole doped system has an extra spin glass phase between $0.02 < x < 0.06$. Above the latter value, superconductivity sets in at low temperatures. With this difference in the spin glass, the phase diagram of the first compound, plotted against abscissae *x,* is a distorted mirror image of the phases in the second.

The highest T_c repeatedly found in the ceramic superconductors is 133 *K* in $HgBa_2Ca_{n-1}Cu_nO_{2n+2}$. This is a member of a group of systems where *n* can vary: $n = 1, 2, 3$ or 4. As *n* decreases T_c declines. As in BSCCO, the copper planes are adjacent, so it appears again that the packing together of CuO_2 planes contributes actively to the superconducting energy gap of the Cooper pairs. Next we consider how the chemistry that we have described influences energy bands for the valence and conduction electrons.

3.3. Hubbard Bands, Valence, and Madelung Potentials

The superconducting compounds that have the highest T_cs and that have been best studied are cuprates [1, 2]. Consider this group in particular, though the general features apply more widely. *Cu* is a 3*d* transition metal, typically divalent but sometimes monovalent. In the ceramic oxide, the 3*d* bands fall into two Hubbard bands: filled and unfilled. Suppose that, in the non-superconducting YBCO compound with $x = 0.5$, the valence band has anionic character, separating the Hubbard bands. After the phase transition that accompanies $x \rightarrow 0$ in YBCO, the consistently valued Madelung potentials [11, Table I, columns 3 and 5] imply holes in the CuO_2 planes, lowering the energies of the anionic bands and raising the energy of the Hubbard bands as in $La_{2-x}Ba_xCuO_{4-y}$ [1, p. 66]. The Madelung potentials are *consistent* when the energy levels are similar for all members on the same ionic species in the unit cell. The levels are determined by firstly the Madelung potentials, secondly by ionization energies on cations, and thirdly by electron affinities on anions. In comparing the various sites, allowance is needed for the different environments in hole planes or reservoirs. Moreover, since the model depends on fine atomic states, and does not account for the band spread that is due to crystal fields, the valence to conduction band gaps are often double those measured by the energy of the fundamental absorption.

The calculations confirm that in YBCO, the oxygen loading acts as a charge reservoir that adjusts the Madelung energies of the atoms in the planes, and also adjusts the Fermi level. The loading sets the environment for holes in the valence band - and therefore for p-type Boson carriers at $T < T_c$. In former times, Madelung potentials were calculated analytically; but more versatile calculations are now, with careful checks for convergence, performed numerically. The most critical part of the calculations for superconductivity describes the energy levels about the planar *Cu* ions. The curvature of the corresponding energy bands determines, in principle, the inverse efective mass of the charge carriers, and therefore the critical temperature T_c. Oxygen loading in the chains has the effect of raising the energies of the conduction electrons that form excitons h^0 on holes formed on the planar *Cu* ions. Naturally, the increase in energy, as x approaches 0, increases the excitonic attraction and curvature of the bands, and therefore raises T_c as in Figure 3.1. Though there is evidence for these effects in the comparison of the two states with $x > 0.5$ and $0 < x < 0.5$ [5, 11], vaguer details imply avenues for future work, both computational and experimental.

The valence bands can in principle be investigated experimentally by microprobe angularly resolved electron photo-emission (μARPES). Kondo et al. [15] for example, observed two dimensional (2D) bands in the planes and quasi one dimensional bands (1D) on the chains. The conduction electrons were well confined within the planes and chains with non-trivial hybridization. As we shall see, interpretation of such data will be enhanced by scanning over various photoelectron energies to find the dispersion in frequency space. The same consideration applies to electron energy band structure calculations, though they are difficult when the unit cells are large, with diverse species. However, localized band structures, such as those calculated by Hu et al. for *Cu* when octahedrally coordinated to *O* [16], provide useful but limited information about dispersion derivatives and band gaps. These molecular orbital type calculations have been used to compare properties due to the different tetrahedral coordination that is found in iron pnictides from the octahedral coordination in the cuprates. ARPES has also been used to map Fermi surfaces: in well doped YBCO the surfaces are cylindrical; in underdoped specimens the surfaces are closed pockets [2, 17].

Meanwhile, carrier densities are measured by the Hall effect, so it is worth discovering whether its measurement in the superconductors correlates with their chemical environment. In HiT_c compounds, the effect can often be interpreted on elementary models in a more or less predictable way as for example in $HgBa_2CaCu_2O_{6+\square}$ [18] and even superconducting YBCO [19]; more interesting are systems that can be processed to have either free electron carriers or hole based carriers, and which we therefore discuss in following section.

3.3. Hall Measurements in YBCO

Apply now this general understanding to the Hall coefficients and resistive data in YBCO measured by Wuyts et al. [17]. Their specimens were thin films, which typically have lower T_cs than bulk specimens, among other characteristic properties that are often different. Nevertheless, electrical contacts can be made more reliably on thin film specimens than on bulk ceramics and this reflects on experimental reliability. The authors report two sets of temperature dependent resistivity data, with $x < 0.5$ and $x > 0.5$, that fell on two straight lines with a change of slope where $x = 0.5$ the stoichiometric value. This is consistent with different carrier types above or below stoichiometry. Moreover their Hall measurements ranged between temperatures T = 100 K to 300K. At these temperatures, $T > T_c$, a component of the resistivities and Hall coefficients respond to holes when $x < 0.5$ and to free electrons otherwise. However, as with *Al*, the *n*-type electrons may suffer negative second derivative in dispersion and so can show positive R_H. Its sign depends in the curvature of the conduction band. This is partly because the superconductors are metallic; in semiconductors, by contrast, the bottom of the conduction band at the forbidden gap must have positive second derivative $\nabla_k^2 \alpha > 0$.

Notice further that though YBCO is superconducting at low temperatures when it is loaded with oxygen; it is otherwise weakly metallic. It is well known that the temperature dependence of the normal resistivity does not follow Drude behavior: pure cuprate superconductors that are optimally loaded, show electrical resistivity ρ that increases linearly with temperature, but that cuts off at temperatures below T_c, so that $\rho(T < T_c) = 0$ in the superconducting state [1 p. 190]. Regressive extension of the linear behavior to lowest temperatures cuts to the origin, $\rho_n(T = 0) \sim 0$. When $x \geq 0.5$, the compound is weakly metallic owing to the *n*-type electrons; but does not superconduct (Figure 3.1).

The measured Hall coefficients were positive both in the set with $x < 0.5$ and with $x > 0.5$. In Figure 3.4 we re-plot $n(R_H(x))$-$n(R_H(0.5))$ from the data of Wuyts et al. Following equation 1, the intrinsic (stoichiometric) carrier density is measured at $x = 0.5$, and since it is subtracted in the modified coefficient $n(R_H)$ in the figure, the value there is zero. More generally, there is a rough inversion symmetry in the figure between the regions $0.3 < x < 0.5$ with $0.7 > x > 0.5$ and this is what should be expected, assuming the dispersion varies slowly with x. The fact that R_H is positive when $x > 0.5$, with *n*-type electrons, is not an inconsistency: the sign on the coefficient depends on the dispersion of conduction bands (appendix A.IV). If these exist in pockets, as measured by ARPES in low-doped YBCO [2, 17], then there is no surprise to find the positive R_H.

Given proper understanding of the Hall effect, physical measurement confirms the conclusions that can be drawn from processing. The charge carriers in the ceramic superconductors are electrons that pass current through holes in their band structures. This occurs, at least, in the normal state, from which the superconducting state must surely be derived. The same holes are capable of bonding Cooper pairs as will be discussed further in the next chapter.

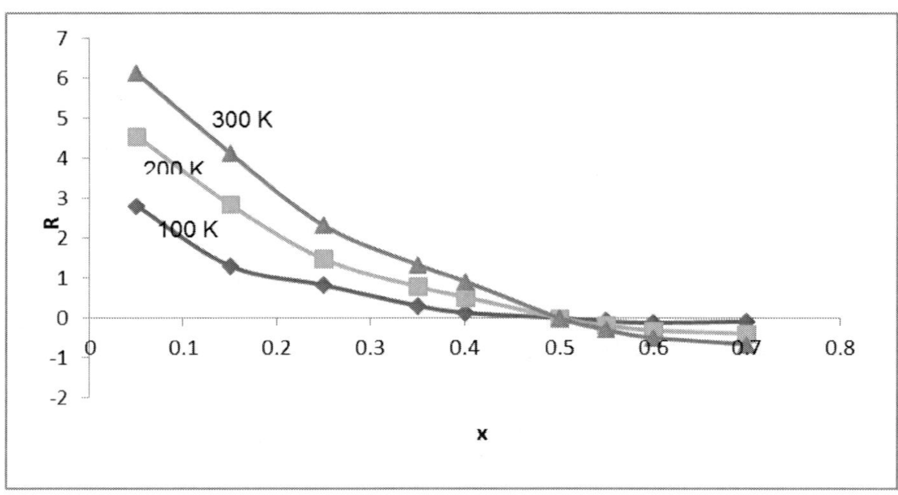

Figure 3.4. Re-plotted values of Hall carrier densities in thin film $YBa_2Cu_3O_{7-x}$ versus x, extracted from the thin film data of Wuyts et al. [17]. Their data are all positive for x < 0.5 to x > 0.5, so we have subtracted an intrinsic component n(x = 0.5) at the stoichiometric condition, the cross from p-type carriers to n-type carriers. Despite a discontinuous slope in resistivity there is a rough inversion symmetry about x = 0.5 of the carrier densities.

REFERENCES

[1] Bourdillon A. and Tan Bourdillon N. X., 1994, *High temperature superconductors, processing and science,* Academic, p. 35.

[2] LeBoeuf D., Doiron-Leyraud N., Levallois J., Daou R., Bonnemaison J. B., Hussey N. E., Balicas L., Ramshaw B. J., Liang R., Bonn D. A., Hardy W. N., Adachi S., Proust C. and Taillefer L. 2007, *Nature* 450 533-536.

[3] Bourdillon A. J., 2015, *Journal of Modern Physics, 2015,* 6 [4] 463-471, DOI:10.4236/jmp.2015.64050.

[4] Bourdillon A. J., *Journal of Modern Physics, 2015,* 6 [14] 2011-2020.

[5] Maple M. B., 1998 https://arxiv.org/pdf/cond-mat/9802202.

[6] Lee P. A., Nagaosa N., and Wen X-G, (2006) *Rev. Mod. Phys* 78 17-8.

[7] Dagotto E., 1994, *Rev. Mod. Phys.* 66 763-840.

[8] Hoffman J. E., 2003, *A search for alternative electronic order in the high temperature superconductor* $Bi_2Sr_2Ca_2Cu_2O_{8+d}$ *by scanning tunneling microscopy, PhD Thesis*, UC Berkeley.

[9] Bardeen J., Cooper L. N. and Schrieffer J. R., 1957, *Phys. Rev.* 108 1175.

[10] Batlogg B., Cava R. J., Jayaraman A., van Dover R. B., Kourouklis G. A., Sunshine S., Murphy D. W., Rupp L. W., Chen H. S., White A., Short K. T., Mujsce A. M. and Rietman E. A., 1987, *Phys. Rev. Lett.* 58 2333.

[11] Tan N. X. and Bourdillon A. J., 1990, *International Journal of Modern Physics B* 4 517-524.

[12] Bourdillon A. J., Beaumont J. H. and Bordas J., 1977, *J. Phys. C. 10* 333-341 (journal later amalgamated to *J. Phys:Condensed Matter* by IOP science).

[13] Acrivos J. V., Chen L., Burch C. M., Metcalf P., Honig J. M., Liu R. S. and Singh K. K., 1994, *Phys. Rev. B* 50 13710-13723.

[14] Bourdillon A. J. 1976, *Spectroscopy of ionic materials using synchrotron radiation*, D. Phil thesis Oxford University.

[15] Kondo T., Khasanov R., Karpinski S., Zhigadlo N. D., Ohta H., Fretwell A. D., Palczewski A. D., Koll J. D., Mesot J., Rotenberg E., Keller H., and Kaminski A. (2006) arXiv:cond-matt/0609034v1 [cond-matt.supr-com] 2 Sep.

[16] Hu J, *Robustness of s-wave pairing symmetry in iron-based superconductors and its implications to fundamentals on magnetically-driven high temperature superconductivity, arXiv:1506. 05791* (2015).

[17] Wuyts B., Moshchalkov V. V. and Bruynseraede Y., 1996, *Phys. Rev. B* 53 9418-9432.

[18] Harris J. M., Wu H, Ong N. P., Meng R. L. and Chu C. W., (1994) *Phys. Rev. B* 50, 3246-3249.

[19] Ando Y, Segawa K, Lavrov A. N. and Komiya S (2002) arXiv:cod-mat/0208096v1.

Chapter 4

EXCITONIC BOSON PAIRS

In previous chapters the nature of the superconducting state was derived empirically from processing data. Clarification was needed: firstly in the concept of negative mass in antiparticles that is modified because of inconsistency with relativity in dispersion dynamics (appendix A.III); and likewise in electrical conduction because of inconsistency with the Lorentz law [ch. 2]. What then is the superconducting state and how does non-resistance occur? In the light of the processing principles, Hall effect measurements are critical, and the extension from chemical holes to bound superconducting pairs is evident and apparent.

The outstanding difference between high temperature superconductors and low temperature superconductors is the sign of the Hall coefficient. Since the Lorentz force acts on particles; not voids nor immobile ions; the experimental positive coefficient is a consequence of the motion of negatively charged electrons in dispersion dynamics, *i.e* on electrons with positive charge/mass ratio, but with negative charge and negative effective mass. In HiT_c compounds, anionic and cationic doping *create holes that substitute for the lattice distortions* that bind Cooper pairs in LoT_c metallic superconductors such as *Nb*. In both types of superconductor, the conventional notion is maintained of antiparallel spins $S = 0$ as in antiferromagnetism, with paired wave vectors, k and $-k$. However, in the ceramics, chemical 'holes' h *are* produced by chemical doping or by unstoichiometric imbalance in ionic charges in the crystallography. In the HiT_c compounds the most significant physical effect is measured in the normal state, namely through positive Hall coefficients. Since the Lorentz force acts on electrons, these must locate on energy bands with negative curvature consistent with negative effective mass. Meanwhile, for conduction at lower temperatures $T < T_c$, the chemical holes are available to bond superconducting Cooper pairs via h^- or h_2^0 excitons. Since they are extended Wannier excitons, not localized Frenkel defects, the two states admix h^- with h_2^0, the latter being the major component. The Cooper pair need not be charged in order to conduct current.

From here options become more numerous and evidence less delineating. Nevertheless, in the process of demystifying physics, some hypotheses are required that are in principle verifiable:

1. In HiT_c, the carriers are electrons in excitonic hole states
2. The Boson pair has a real wave function with antinodes that overlap periodic holes (appendix A.I). In dispersion dynamics the superposition of a particle with its antiparticle wave produces a real wave function. This becomes an electromagnetic wave in annihilation [1]. The solutions to the equations of relativity make the particle right handed *RH* and the antiparticle left handed *LH*. In HiT_c, by contrast, both electrons are right handed but their effective mass is negative. We propose that in the HiT_c superconducting state, electron waves of opposite momenta, but the same right handedness, superpose to a real trigonometric wave function. This is the consequence of superposition of a Cooper pair with wave vectors k and $-k$ having angular frequencies $-\omega$ and ω respectively. Though the purely spatial parts are contiguous (figure A.I.1), when the temporal part is applied the result is like the particle-antiparticle image in figure A.I.2. The superposition is a real wave function with net momentum zero ($\Sigma k_i = 0$). It can neither create phonons nor scatter on the lattice. This offers no

electrical resistance, and is consistent with the Meissner-Ochsenfeld effect. The exciton h_2^0 carries zero net current.

3. Typically the holes occur at cell boundaries on Cu^{2+} ions. So long as their temperature is less than critical ($T < T_c = \varepsilon_g/k_B$, the superconducting gap energy divided by the Boltzmann constant), the wave representing the Boson pair, transports electric current at the phase velocity $v_p = \omega/k$. The superconducting pair state is energetically stable like a bound atomic wave function, but strung out across the lattice. At electrodes, Cooper pairs decompose and reform by Bose Einstein statistics, with net current transport. This decomposition and reformation of the Cooper pair applies equally to conductivity in LoT_c alloys.

The similarity that occured in dispersion dynamics between antiparticles and the Boson pair, correlates with equation (1.1). There, $\phi^*(k,\omega) = \phi(-k,-\omega)$ ignoring the spin function appendage. The complex conjugate reverses the propagation direction.

The first question that arises from this hypothesis for the superconducting state in HiT_c is, "How can it be tested?" Some tests for this theory are already evident. They are the many inconsistencies in competing theories, in particular: with regard to positive mass in antiparticles; unphysical implications in the wave dynamics of the antiparticle when $|k| = |m|$; contradictory application of the Lorentz law in the Hall effect to non-existent, positively charged particles; general application of wave dynamics in other applications of physical theory; etc. Further tests and confirmation may well be found in extended computations of Madelung energies in the ionic, ceramic superconductors, and by extended and focused measurements of band structures, in spite of the difficulties involved. The most rewarding test might be a discovery – by application of the principles involved - of new systems with higher T_cs than those already known.

References

[1] Bourdillon AJ, *Journal of Modern Physics,* **5**, [1] 23-28 (2014) DOI: 10.4236/jmp.2014.51003.

[2] Bourdillon AJ, *Journal of Modern Physics* **8** 483-499 (2017) 11DOI: https://doi.org/10.4236/jmp.2017.84031.

APPENDICES

APPENDIX A.I. HANDEDNESS IN THE STABLE WAVE PACKET

By mathematical convention, equation 1.1 is right handed. This is easily seen when the argument has, $X > 0$, k > 0. In Figure A.I.1, the two corkscrews are right handed, and their forward axes represent their propagation directions, *i.e.* parallel to the phase velocity $v_{p,}$ and typically also the group velocity v_g.

Superposing the two wave functions $\phi(k,\omega)$ and $\phi(-k,-\omega)$ produces the identical carrier wave in space, but with two components travelling in opposite directions v_g and $-v_g$.

While the function $\exp(iX)$ is right handed (RH), as in the expansion $\cos(X) + i.\sin(X)$; $\exp(-iX)$ is left handed (LH) as in $\cos(X) - i.\sin(X)$. In dispersion dynamics the electron is right handed; the antiparticle left handed like the mirror images in the figure.

Consider complex chirality of the wave function in equation 1.1, and ignore magnetic spin as an appendage whether the two are in fact interdependent or not. Our major concern is for electrostatic interactions between conduction electrons. This is because the Cooper pair bonding is supposed electrostatic, like crystal bonding, whether in antiferromagnetics or whatever. Figure A.I.2 contrasts the chiralities of a right handed electron with its left handed, mirror image, antiparticle. The *LH* handedness of the antiparticle is a consequence of its negative energy. The wave vector of the antielectron is antiparallel to its direction of propagation, given by the group velocity vector v_g. Superposition of the two waves ϕ_k and ϕ_{-k}^* results in a real wave function with cancellation of the imaginary part of the individual Fermionic functions. Superposing the two waves produces a real wave because the imaginary parts cancel. In annihilation, the real wave transforms to a photon (or two photons); the Cooper pair to two electrons.

In dispersion dynamics, two RH, electronic, Bloch waves, with opposing wave vectors, form a real wave function: $\exp(iX) + \exp(-iX) = 2\cos(X)$. This has nodes and antinodes. The antinodes are attracted to periodic holes in the superconductor lattice. The mean wave vector of the Cooper pair is zero so that it cannot excite phonons or scatter off the lattice. Charge transport is therefore resistanceless and occurs by break-up of the pair at electrodes. Consistent with Bose-Einstein statistics, the Cooper pairs may reform anywhere across the lattice. The current transport occurs with the phase velocity, faster than the speed of light.

The photon is a Boson like the Cooper pair, but is formed from two real waves, electric and magnetic, that are mutually out of phase by $\pi/2$. Photons can superpose in phase, as in a laser beam and also when two photons have opposite directionality, ϕ_k and ϕ_{-k}, as in the laser cavity.

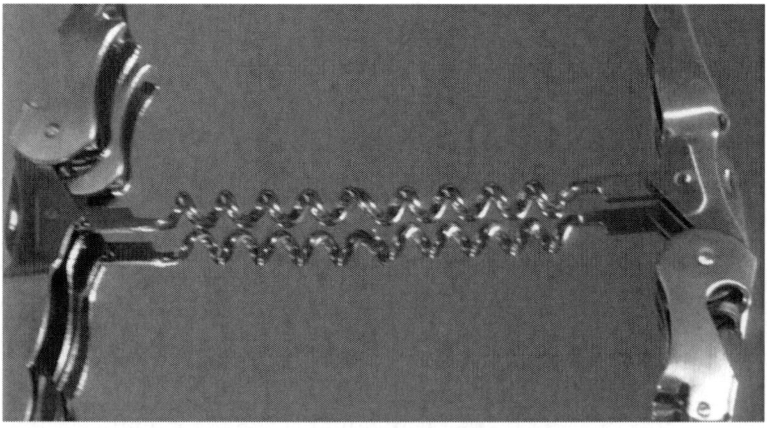

Figure A.I.1. Two *RH* corkscrews end-to-end ($k_1 = -k_2$) in a continuous wave, with the corresponding left handed image below (with mirror plane $M//x$).

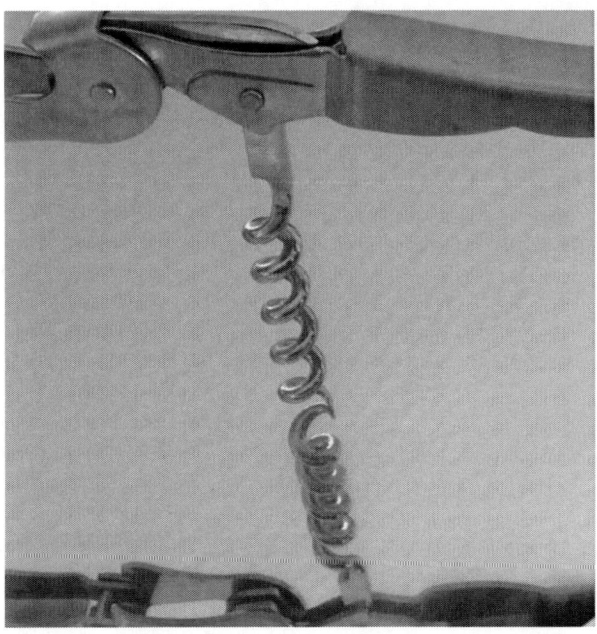

Figure A.I.2. RH corkscrew with discontinuous mirror image ($M\downarrow x$) and reversal of chirality.

In dispersion dynamics, there is a speculative reason why Bosons are distinguishable, *i.e.* due to easy termination at antinodes. Termination of the wave packet is a requirement for quantization. By contrast, the complex termination of Fermions makes them indistinguishable: requiring joint wave functions that are anti-symmetric with respect to exchange, as required by Pauli exclusion.

Further evidence for the real wave function lies in the facts of the dc and ac Josephson effects [1, 2, 3] at a junction. Fermions and Bosons both tunnel, but the phase of superconducting pairs, is retained on both sides of the junction. This occurs because the phase is not disturbed by the junction at a wave node.

Moreover, helicity takes account of the reference frame of the observer. Chirality is a property of a particle. The helicity of a particle with mass m travelling with velocity $\boldsymbol{v}_g$, would be observed to change sign when the observer accelerates from a velocity $v < v_p$ to $v > v_p$. In dispersion dynamics, since, in a particle $v_p > c$, the helicity change is not observable.

REFERENCES

[1] Bourdillon A and Tan-Bourdillon NX, *High Temperature Superconductors: Processing and Science*, Academic Press, N. Y., 1994, ISBN 0-12-117680-0.

[2] Tilley DR and Tilley J, *Superfluidity and superconductivity, 1986, Hilger ISBN 0-85274-807-8.*

[3] Lin JY, Gurvitch M, Tolpygo SK, Bourdillon A, Y. Hou SY and Phillips JM, *Phys. Rev. B,* 54 R12717- 20 (1996).

APPENDIX A.II. UNPHYSICAL SINGULARITY FOR A NEGATIVE EIGENVALUE WITH POSITIVE KINETIC ENERGY IN THE ANTIPARTICLE

The wave functions for the stable wave packet conflict with Dirac's model for the positron. The conventional model has a negative eigenvalue for the first order Dirac equation; with positive mass; positive kinetic energy; and positive charge. Several of these properties conflict with the stable wave packet (eq. 1.1) in dispersion dynamics, after insertion in the second order Einstein equation for relativity (eq. 1.2). Firstly, if the eigenvalue is negative, the mass should be negative, because in the rest frame, the energy is the mass; secondly if the mass is negative the kinetic energy must be negative, because a positive kinetic energy results in singularities in the group velocity and phase velocity of the antiparticle when $|k| = |m|$ (Figures A.III.1 and A.III.2). The singularities are unphysical (infinite mass and energy when $v_g = c$) and are not observed. The wave vector must be negative. So also must the angular frequency be negative, otherwise the wave packet will self-annihilate. More debatable is the charge to mass ratio e/m [1], a physical constant.

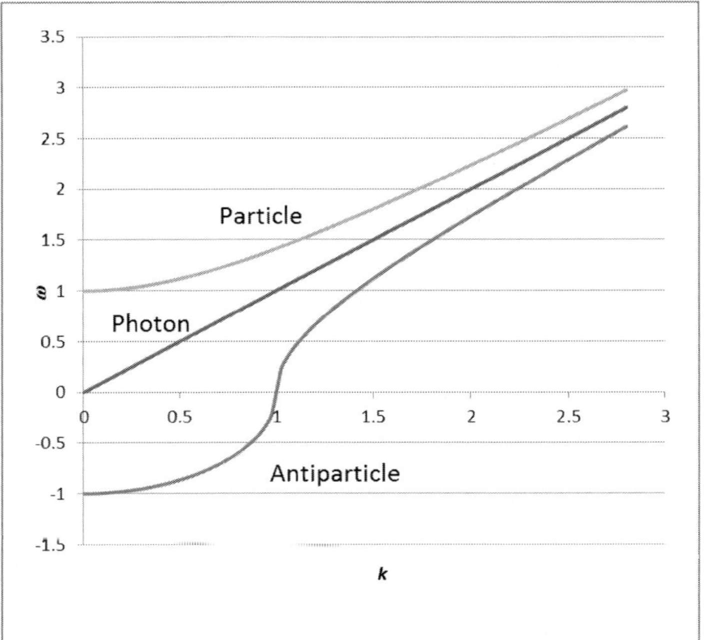

Figure A.II.1. Angular frequencies (energy) calculated from equations 1.1 and 1.2 for a particle with mass $m = 1$; and values for an antiparticle with negative rest energy in unit -1, supposing *positive kinetic energy*. When $|k| = |m|$, the energy is singular. Calculations are made in simplified units. Consistency requires all of E, m,ω and k to be negative, along with kinetic energy in the antiparticle.

The group and phase velocities of the particle and antiparticle are sufficiently fundamental that their conflict with the conventional model cannot be ignored. Since the mass of the antiparticle is negative, consistency with electrostatic, magnetic and electromagnetic equations require that the charge on the antiparticle should be negative like the electron. After taking into account negative mass in Newton's first law of motion, it turns out then that charge conjugation symmetry is replaced by M' symmetry, where M' denotes negative masses in antiparticles [1]. In dispersion dynamics, the positron is replaced by the negatron. For example, where conventional formulae had:

$$ma = (\mp e)E \quad \text{(AII.1)}$$

for the acceleration a of a particle with mass m and charge $-e$ for the electron, or $+e$ for the positron, in an electric field E; this is replaced in dispersion dynamics by:

$$(\pm m)a = -eE \tag{AII.2}$$

where the acceleration properly reverses for the antiparticle in both equations. Similar changes occur in equations for the Lorentz law and in electron-antiparticle interactions.

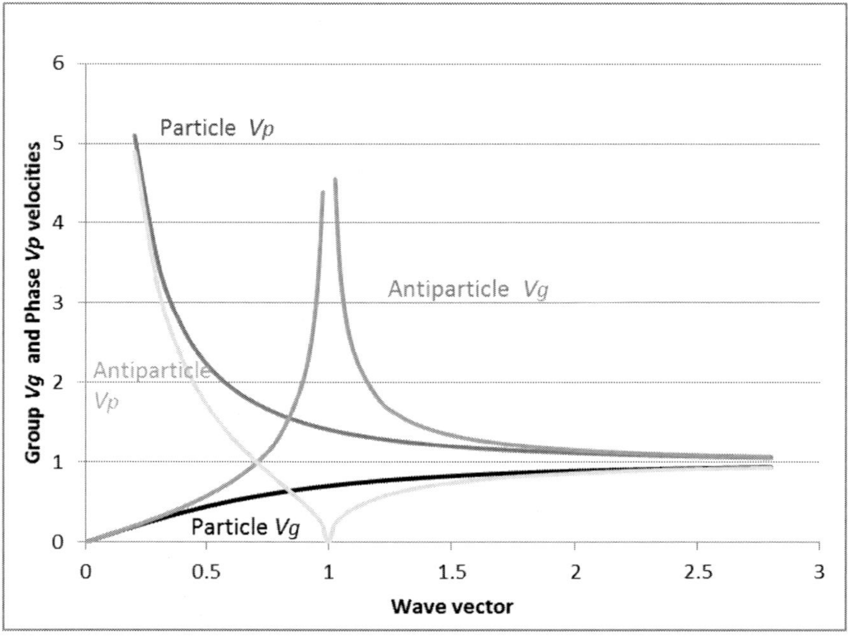

Figure A.II.2. Group v_g and Phase v_p velocities for a particle, mass = 1 (darker traces), calculated from equations 1.1 and 1.2; and *erroneous velocities* for an antiparticle, mass = -1 (lighter traces). The singularities calculated for the antiparticle when $k = m$ are not observed and are not physical, nor are the group velocities $v_g > c = 1$. Where the rest mass energy is negative, the kinetic energy cannot be positive. Units are simplified.

The negative charge with negative mass reacts in the same way as a positive charge with positive mass — but with important differences that allows for the consistency and added features (some needed in this book) of dispersion dynamics. In consequence CPT symmetry (charge conjugation, parity, time [2]) is replaced by M'PT symmetry in all physical interactions, but excepting the weak nuclear force [3]. There are further important differences and perhaps the most notable is Dirac's notion of filled states of antiparticles which become holes on antiparticle creation. Our antiparticles gain gravitational force when they are created; his holes are not needed.

REFERENCES

[1] Bourdillon A. J., 2014, *Journal of Modern Physics,* 5 23-28 doi: 10.4236/jmp.2014.51003.

[2] Villata M, 2011, *EPL* 94 20001 doi: 10. 1209-5075/ 0295/94/20001.

[3] Wu C. S., 1956, *Phys. Rev. 104* 254, doi:10.1103/PhysRev.104.254.

APPENDIX A.III. VARIOUS CONSEQUENCES IN DISPERSION DYNAMICS

The stable wave packet is sufficiently fundamental that many adjustments are associated: some are conventional and optional; others have wider reach. Two examples are described as loose ends that have to be tied. In particular, negative mass has cosmological implications that are more surely not within the scope of this book, for example in annihilating galaxies of alternative masses, which may either mutually attract or repel[12]. In the laboratory, experimental measurements on these forces have been recently described [1].

[12] Negative mass does not explain constant rotational velocities in spiral galaxies; this is easily calculated to be a conventional, chain-like, many-body effect of gravity.

A.III.1. Dynamics with Negative Energy and Mass

Since equations 1.1-1.6 in chapter 1 require solutions in which antiparticle mass and kinetic energy are negative together, annihilation and creation events can be represented conveniently in novel form. This contains an economy in representation by either the rest frame, or in a moving frame. If the photon energy and momentum causing the event is given by the diagonal within the rectangle in Figure A.III.1, conservation of energy is defined by the rectangle, then momentum is conserved about the center of mass. This is maintained by transporting the center of the rectangle to the origin of the graph. In a moving frame, respective energy and momentum for the particle and antiparticle are then represented by arrows shown: upwards for the free particle; downwards for the free antiparticle onto its energy band. In the center of mass frame, the net momenta of the particle and antiparticle are zero; the momentum of the photon is balanced by collision with a third particle. Of course in a particle accelerator, negative energy applied to negative mass requires positive engineering delivery.

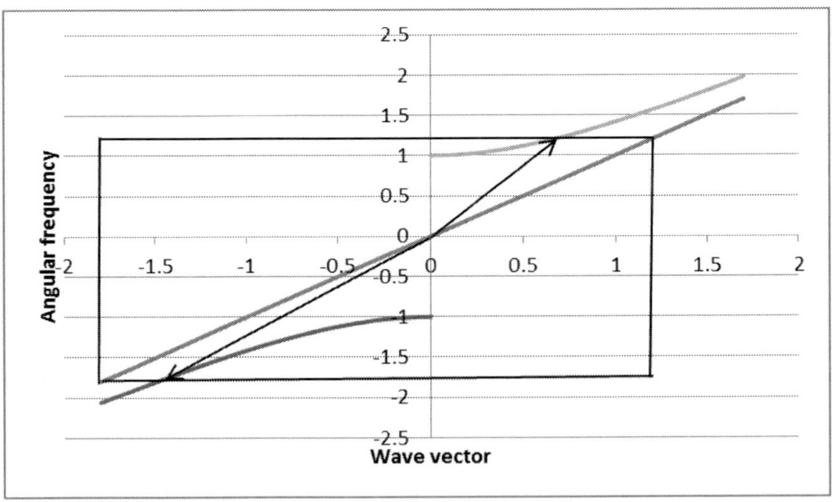

Figure A.III.1. Energy conserved in a construction using negative mass energy and negative kinetic energy in an antiparticle (ch. 1, Figure 2). Conservation of momentum requires collision with a third particle.

A.III.2. Electromagnetic Energy in 5-Dimensions

So far the discussion has been of free particles. Assume, consistent with the Dirac relativistic equation with the relativistic Klein-Gordon equation, that the dynamics are described in five dimensions that include mass as well as the direction of propagation; time; and two transverse directions. Rest mass is discrete owing to quantization in nuclear structure in the standard model; the other dimensions vary continuously. Now, include acceleration in an electromagnetic field and leave nuclear forces and gravitation for another time. In order to plot the constraints that determine the travelling wave group in relativistic frames, first summarize well known constructions and then illustrate them graphically.

Equations 1.1 and 1.2 in chapter I can be made relativistic by writing $X = g_{\mu\nu}k^{\mu}x^{\nu}$ in relativistic form using the Einstein summation convention for repeated indices; where $k^{\mu} = (k^1, k^2, k^3, k^0)$ represents the three covariant components of wave vector plus angular frequency; where $x^{\mu} = (x^1, x^2, x^3, x^0)$ represents corresponding covariant components of space and time; and where $g_{\mu\nu}$ represents the metric tensor. To include mass, the four dimensional mathematics can easily be increased to five dimensions that include mass by adding a fifth coordinate, and with the metric tensor, $g_{55} = 1$, *i.e.* with the opposite sign to g_{44} in the convention used by Ziman [2]. The addition could be useful in describing nuclear interactions, but possible applications are not developed here. To see how the formulae for the travelling wave group are used, consider first free particles and then accelerated particles.

Notice that the classical Hamiltonian for a free particle, $H = p^2/2m$, which is used in its derivative equivalent, $H = -i\hbar/2m.\ \nabla^2\psi$ in the Schrödinger equation, is approximate. The Hamiltonian is misleading when compared to the proper relativistic formula, including mass. This is illustrated in figure A.III.2 for a free particle. Here, equations 1.2 and 1.3 are represented by the theorem of Pythagoras considering, in the first instance, positive values $m^2 = (\omega + k)(\omega - k)$. The vertical (dashed) axis through k = 0 contains the discrete rest mass; the horizontal axis on k contains the locus of reference frames (RF) that can be used to observe the motion. Angular frequency ω is constrained by relativity. The oblique hatched area is equal to the vertical hatched area, m^2. Inverting through the origin of the figure, corresponding relationships between negative ω, k and m in antiparticles can be represented. On this general base, consider the application of external electromagnetic fields.

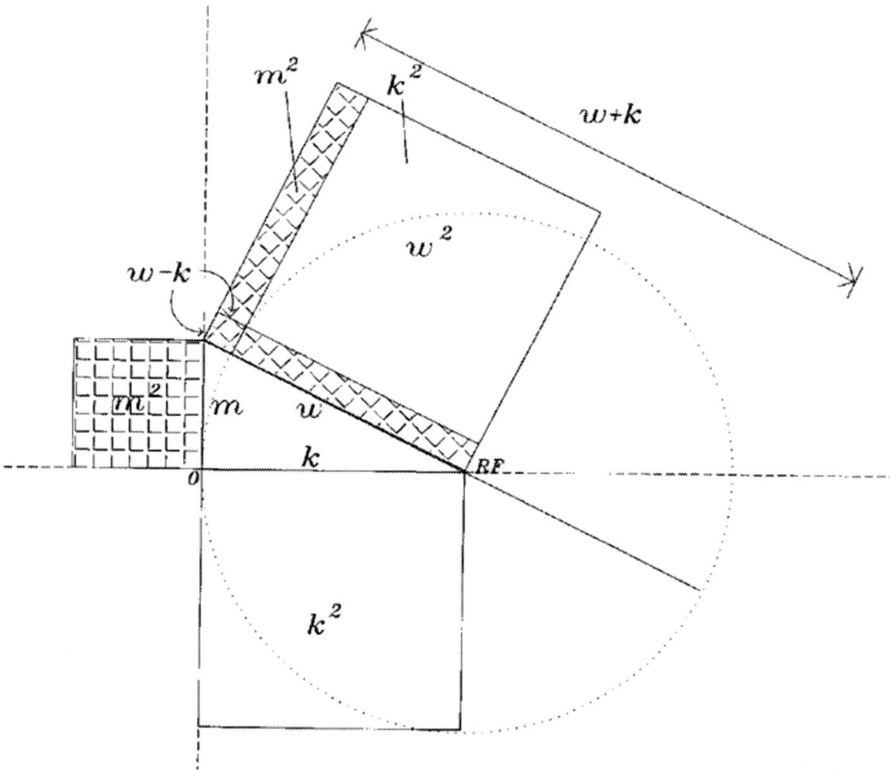

Figure A.III.2. Illustration of Pythagorean relationships in eq. 4 for a free particle at particular reference frame RF on the horizontal axis, when $m = \sqrt{(\omega + k)(\omega - k)}$ on the vertical axis. The squared mass m^2 is shown hatched first vertically, and secondly obliquely. The horizontal axis is the locus of RF.

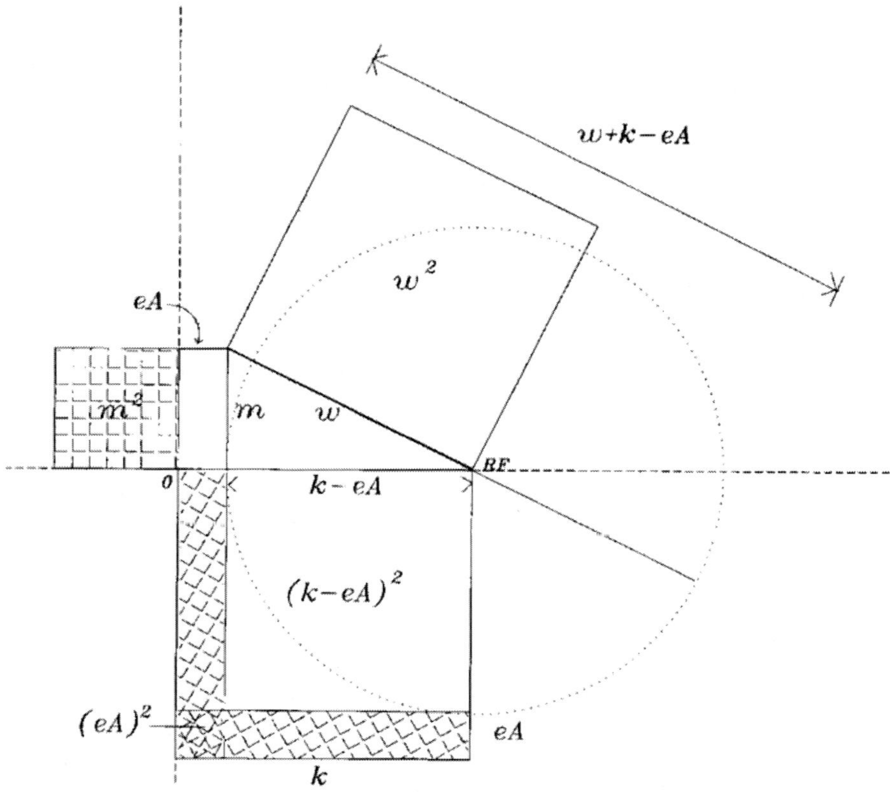

Figure A.III.3. Illustration of Hamiltonian terms for a particle in an electromagnetic magnetic potential eA. The square of mass m^2 is shown hatched, first vertically and secondly obliquely. Notice the double hatched square of area (eA)^2 which is subtracted twice, within e$k.A$, from k^{2,} while being added once in equation AII.2.1. The term (eA)^2 is therefore not duplicated in the diagram.

The simplest way to represent the dynamics due to these fields is given by classical analogues [3]. This allows us to replace the momentum p by p-eA, where e is the electric charge on the particle and A represents the four dimensional magnetic vector potential with scalar potential. The Hamiltonian then includes terms in (p-eA)^2 and in m^2. Expanding these,

$$\omega^2 = g_{\mu\nu}k^\mu k^\nu - 2eg_{\mu\nu}p^\mu A^\nu + e^2 g_{\mu\nu}A^\mu A^\nu + m^2 . \qquad \text{(A.III.1)}$$

Figure A.III.3 shows how the four terms, expressed in our units, are each accounted in Pythogorean constraints on ω and k. These figures show how the travelling wave group may be used consistently with relativistic equations. Notice that when $|k| < |eA|$, the model describes a bound state, as in the application of the Schrödinger equation to atomic orbitals in sub-relativistic approximation. Moreover, the formalism gives a more general solution than does the elementary application of magnetic interaction with spin, as in the inner product of the vectors $\boldsymbol{\sigma}.\boldsymbol{B}$. As is conventional for simplicity, that spin is sometimes treated non-relativistically, as a separable variable on the wave function. The Dirac four-vector explanation for spin in the relativistic wave function will then give validity to a spinor matrix in our travelling wave group with inversion. The first order illustration of an electromagnetic solution for the second order relativistic equation is shown in Figure A.III.3.

References

[1] Khamehchi MA, Khalid Hossain, Mossman M. E., Yongping Zhang, Busch Th., McNeill Forbes M. and Engels P., 2017, *Phys. Rev. Lett* 118 155301.

[2] Ziman JM, Elements of advanced quantum theory, 1969, Cambridge.

[3] Bourdillon AJ, *Journal of Modern Physics,* 4 705-711 (*2013),* doi: 10.4236/jmp.2013.46097 http://www.scirp.org/journal/jmp.

Appendix A.IV. Comprehension and Ramification

The book began with a comparison between the dynamics of a free electron (Figure 1.3) with the behavior of an electron in a crystal field. The former are simpler. The dispersive gradient of the free particle is always positive, $\nabla_k \omega > 0$, so also therefore is the group velocity $\boldsymbol{v}_g$ that defines the propagation direction. This is true for the free electron as for its antiparticle. The two most significant differences between the electron and its antiparticle are the *rest mass* which, in dispersion dynamics, is negative for the antiparticle, and the *second derivative* for the dispersion which is therefore also negative $\nabla^2_{k_x} \omega < 0$ (equation 1.20). A third difference is the e/m ratio which is negative for the electron and positive for its antiparticle. Antiparticle behavior was used to exemplify the Hall effect in metals, semiconductors and ceramic superconductors. Actually the dispersion that is calculated in such solids is more complicated than the dispersion of more easily understood free particles. The discussion indicated the general importance of less strongly bound, outer, group I electrons in the conductivity of metals and semiconductors. The configurational s-type electrons are free electron like with positive gradients in dispersion and positive second derivatives. In semiconductors the forbidden gap modifies those features; though in insulators the outer group I electrons lose significance owing to covalent or ionic bonding.

Despite these trends, the gradient at the base of the conduction band for any specimen may be positive or negative and directional. Meanwhile, the second derivative may also be calculated to have a range of positive or negative values in the various materials. The free electron dynamics therefore have significant value as an aid to comprehension, especially as a guide to general trends; but the dynamics are limited by their actual scope of application.

Added to this complexity is the fact that the Lorentz force in the superconductor under Hall measurement, acts on electrons whose v_g may, despite general trends, be positive or negative and whose mass may also be positive or negative. Both of these properties depend on details of the dispersion at the base of the conduction band. Moreover the relations may be different on the x-axis (Figure 2.1) from those on the z-axis. However, in the low magnetic field approximation, the field dependent electric current, qv_x, is always positive. The Lorentz force is therefore independent of the sign on e^2. The Hall coefficient depends on effective mass, and therefore on the second derivative in the dispersion (equation 1.20). For the four cases involving positive or negative carrier charges, the sign of the Hall coefficient can be tabulated in dispersion dynamics (Table A.IV.1). Free particles are restricted to the first row. In solids, any of the four values may, in principle, occur, depending on their individual band structures and type of carrier. The cuprate superconductors belong to the second row.

Experimental measurements of Hall coefficients in the cuprate superconductors therefore confirm the importance of holes in the normal conductivity. However, there are limits to the reliability of band structure calculations for these multi-element compounds, if they are even realistic at present. Moreover experimental confirmation of the detailed theory is itself problematic. Perfect compliance with the ideal (free electron) model should not be expected, even though the strongest empirical evidence exists for excitonic bonding of Cooper pairs in the superconductivity.

Table A.IV.1. In dispersion dynamics, the sign on R_H depends on the carrier drift and on the mean value for the dispersive second derivative at the base of the conduction band.

	$m_{eff}^{-1} = \nabla_k^2 \omega$	**>0**	**<0**
v_g			
> 0		R_H>0	<0
< 0		<0	>0

INDEX

A

B

C

D

N

O

P

R

S

T

U

V

W

Y